A Quick Guide to Welding and Weld Inspection

A Quick Guide to
Welding and Weld Inspection

Steven E. Hughes

Series editor: Clifford Matthews

Matthews Engineering Training Limited
www.matthews-training.co.uk

WOODHEAD PUBLISHING LIMITED

Oxford Cambridge New Delhi

Published by Woodhead Publishing Limited, 80 High Street, Sawston, Cambridge CB22 3HJ
www.woodheadpublishing.com
and
Matthews Engineering Training Limited
www.matthews-training.co.uk

Woodhead Publishing India Private Limited, G-2, Vardaan House, 7/28 Ansari Road, Daryaganj, New Delhi - 110002, India
www.woodheadpublishingindia.com

Published in North America by the American Society of Mechanical Engineers (ASME), Three Park Avenue, New York, NY 10016-5990, USA
www.asme.org

First published 2009, Woodhead Publishing Limited and Matthews Engineering Training Limited

British Library Cataloguing in Publication Data
A catalogue record for this book is available from the British Library.

Library of Congress Cataloging in Publication Data
A catalog record for this book is available from the Library of Congress.

Woodhead Publishing ISBN 978-1-84569-641-2 (book)
Woodhead Publishing ISBN 978-1-84569-767-9 (e-book)
ASME ISBN 978-0-7918-5950-6
ASME Order No. 859506
ASME Order No. 85950Q (e-book)

Typeset by Data Standards Ltd, Frome, Somerset, UK

Contents

Foreword by Series Editor

We are pleased to introduce this new book by Steve Hughes in our ongoing *Quick Guide* series. Intended for inspection engineers and technicians the *Quick Guide to Welding and Weld Inspection* will provide a useful reference for those involved in the plant inspection discipline.

Steve Hughes provides valuable lecturing input to our ASME Plant Inspector and API 510/570/653 Certified inspection training courses run by Matthews Engineering Training Ltd, bringing his technical knowledge underpinned by sound engineering practical experience. As a discipline, Plant Inspection incorporates many subjects, but welding and weld inspection forms a common thread running through the majority of them. For pressure systems in particular, weld joining design, welding techniques, NDE and defect acceptance are prime considerations in equipment repair and assessing the risk of existing equipment.

The documentation aspects of weld inspection often remain a bit of a mystery to plant inspectors (even experienced ones). In assessing the quality of weld repairs it becomes all too easy to assume that the necessary WPS/PQR weld documentation and welder qualifications are in place, without properly checking. Equally it is easy to be intimidated by the technical detail and terminology symbols and acronyms. You should find this book useful in helping you through such situations.

Despite the predominance given in technical publications to advanced technology-intensive NDE and inspection activities, it remains a fact that 90% of all pressure equipment inspections include only well-proven established inspection activities. In writing this book, Steve Hughes has addressed these real issues of most practical weld inspections, distilling a wide subject into a simplified and digestible format. You don't have to be a qualified welding engineer or metallurgist to read this book, and you may still benefit from it if you are.

Finally, we are always interested in hearing from people who are interested in writing a *Quick Guide* book (like this one) in an inspection-related subject. You don't have to be an experienced author or an acknowledged expert on the subject (all experts are self-appointed anyway – we know that), just have sufficient experience to know what you are talking about and be able to write it down in a way that other people will understand. We'll do the rest. Contact me on enquires@matthews-training.co.uk or through our website www.matthews-training.co.uk.

Cliff Matthews
Series Editor

Introduction

This book was written to provide a quick guide to welding inspection which is easy to read and understand. There are many books covering all aspects of welding (many of them go into great detail on the subject) but it is difficult to find books specifically covering weld inspection requirements. This book's subjects are purposely not covered in great detail because it is assumed that the reader is able to find detailed books on specific subject areas of particular interest. What this book will do is give you a basic understanding of the subject and so help you decide if you need to look further. In many cases the depth of knowledge required for any particular welding-related subject will be dependent on specific industry requirements. In all situations, however, the welding inspector's role is to ensure that welds have been produced and tested in accordance with the correct code specified procedures and that they are code compliant. Code compliance in this sense means that the weld meets all the requirements of the defect acceptance criteria specified within the code.

Inspectors considering training to achieve certified welding inspector status under certification schemes such as CSWIP (Certification Scheme for Welding Inspection Personnel) or PCN (Personal Certification Number) will find the book a useful pre-course learning aid giving coverage of the 'body of knowledge' they are expected to be familiar with. Non-welding personnel will find it a useful introduction to the world of welding inspection. Some people believe that a welding inspector must have previous welding experience, but this is not necessarily true as welding and welding inspection are two totally different subjects. Welding is naturally a mainly practical 'skill of hand' process and requires dexterity and good hand-to-eye coordination from the welder. The inspector does not require this practical skill but must be able to oversee the welding process, take accurate measurements, interpret the requirements of codes and standards, and ensure that completed welds are in compliance with the relevant code requirements. A good inspector is one who does not take shortcuts and ensures that procedures are properly followed.

S. E. Hughes
Author

A 10 Minute Guide

Here are some frequently asked questions about welding inspection subjects. You can answer them if you spend ten minutes reading the following information.

1. What is the difference between the main welding processes? (four minutes)

MMA has an arc struck between a consumable flux-coated electrode and the workpiece. The electrode melts and fills the joint with weld metal. The flux coating melts and produces gas to shield the arc. The melted flux also helps to remove impurities from the weld and forms a layer of slag on top of the weld. This slag must be removed between runs or it can cause slag inclusions within the weld. This is the most commonly used outdoor site welding process.

MIG/MAG has an arc struck between a reel-fed consumable solid wire electrode and the workpiece. It does not produce slag because the arc is shielded by an inert gas (active gas for MAG). It deposits weld metal quickly and can be used semi-automatically, mechanized or automated. This process can achieve fast weld metal deposition rates.

FCAW is similar to MIG/MAG but the consumable reel-fed electrode is hollow with a flux contained inside. The arc can be self-shielded using only the melting flux or a secondary gas shield can be used. Slag will be produced and must be removed between runs. Basically, this process gives a combination of the advantages (and disadvantages) of MMA and MIG/MAG.

TIG has an arc struck between a non-consumable tungsten electrode and the workpiece. Filler in wire or rod form is added separately. It does not produce slag because the arc is shielded by an inert gas. It is a very slow process but produces very high quality welds.

SAW has an arc struck between a reel-fed consumable solid wire electrode and the workpiece. The arc is formed underneath a layer of flux and is therefore not visible to the operator. This is a deeply penetrating process requiring good penetration control. It has very fast deposition rates but is generally restricted to welding in the flat or horizontal–vertical positions (otherwise the flux would fall out).

Spend four minutes reading through the welding processes, noting the differences and making sure you recognise the acronyms used for each process.

2. Why use welding symbols? (one minute)

Construction drawings need to pass on information clearly and within a limited space. Welding information is passed on by a method of conventions and symbols.

Spend one minute looking over the different symbols for butt welds and fillet welds.

3. What are codes and standards? (one minute)

Construction codes and application standards contain the rules that must be followed when providing a specific product or service. They contain information on design, manufacturing method, acceptable materials, workmanship, testing requirements and acceptable imperfection levels. They do not contain all the relevant data required for the design, manufacture, testing and inspection but will reference other standards and documents as required.

Spend one minute reading the definitions of codes and standards.

4. What is welding procedure qualification? (one minute)

Welding procedure qualification is carried out to prove that a welded joint meets the mechanical, metallurgical and physical properties required by a code or specification. It also enables repeatability by encouraging a systematic approach. The welding procedure qualification documents consist of a Welding Procedure Specification (WPS) and a Procedure Qualification Record (PQR) containing essential and non-essential variables.

Spend one minute reading about essential, non-essential and supplementary essential variables.

5. Why does a welder need qualification? (one minute)

Welders require qualification to prove that they have sufficient knowledge and skill to produce a weld in accordance with a welding procedure and achieve a result that meets the relevant specification. A welder qualification test form records the range of variables qualified from the essential variables.

Spend one minute looking at the variables contained in a WPQ form.

6. What are welding defects? (two minutes)

There is no such thing as a perfect weld because all welds contain imperfections of some sort. These imperfections need to be assessed to determine if they will have a detrimental effect on the welded joint. They are normally assessed against the acceptance criteria specified in the relevant code or standard but only an imperfection found to be outside the acceptance criteria limits will be classed as a defect and require action to be taken. The action to be taken could include:

- rejection of the component;
- removal of the defect and re-welding of the joint;
- a fitness-for-purpose analysis being carried out with a concession being granted allowing the defect to remain.

Spend two minutes determining what defects can occur in welds.

Chapter 1

Abbreviations, Terminology and Welding Symbols

Abbreviations

It is useful to have an awareness of abbreviations used within the world of welding and welding inspection. Some commonly used ones are:

API:	American Petroleum Institute
ASME:	American Society of Mechanical Engineers
BM:	base metal
BS:	British Standard
BSI:	British Standards Institute
CVI:	close visual inspection
DT:	destructive testing
EN:	European standard (Euro Norm)
GVI:	general visual inspection
HAZ:	heat affected zone
ISO:	International Standards Organisation
LCT:	lower critical temperature
MT/MPI:	magnetic testing/magnetic particle inspection
NDE/NDT:	non-destructive examination/non-destructive testing
PQR:	Procedure Qualification Record
PT/LPE:	penetrant testing/liquid penetrant examination
PWHT:	post-weld heat treatment
RT/RAD:	radiographic testing/radiography
SWI:	senior welding inspector
UCT:	upper critical temperature
UT:	ultrasonic testing
UTS:	ultimate tensile strength
VI:	visual inspection
VWI:	visual welding inspector

WI:	welding inspector
WM:	weld metal
WPS:	welding procedure specification

Common terms

Longitudinal direction:	along the length of the weldment (parallel to the weld)
Transverse direction:	along the width of the weldment (perpendicular to the weld)
Short transverse direction:	through the weldment thickness
Weldment:	the combined weld, HAZ and base metal
Fusion weld:	a weld produced by joining materials in a molten state
Yield point:	the point at which plastic deformation starts in a material

Joint terminology

Before welding takes place the parts to be joined must be prepared and arranged into the required form. The most common forms of joint are butt (or groove) joints, T joints and lap joints. Butt joint members are fitted edge to edge, T joints are fitted edge to surface and lap joints are fitted surface to surface (see Fig. 1.1).

Weld terminology

Types of weld used on the joints are butt (or groove) welds, fillet welds, edge welds, plug welds and spot welds. The type

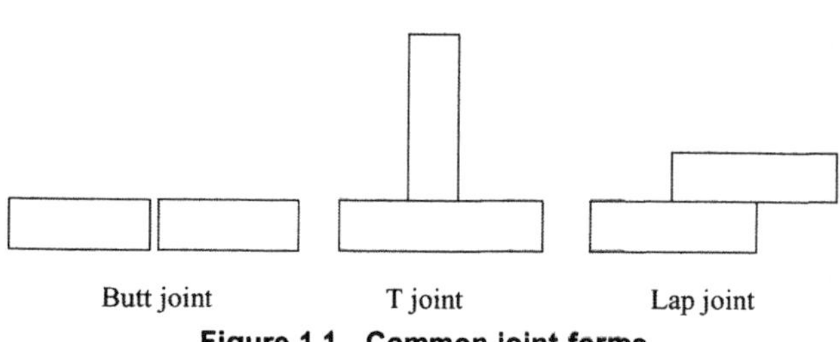

Figure 1.1 Common joint forms

of weld will be determined by the joint set-up but the most commonly used welds are butt or fillet welds.

Full penetration butt welds are normally the strongest type of weld with the strength contained within the throat of the weld, as indicated in Fig. 1.2(a). The throat is equal to the thinnest of the parent materials to be joined.

Fillet welds also contain their strength in the throat and the design throat size is normally (but not always) based on the leg size, as shown in Fig. 1.2(b). In visual inspections we can normally work out the design throat size of a mitre fillet with equal leg lengths by multiplying the leg length by 0.707. Conversely, the leg length can be calculated by multiplying the design throat by 1.414.

Butt weld joint preparation consists of preparing both edges and then arranging them together to permit the required depth of weld penetration to be achieved. Full penetration welds are the most common but many applications will only require partial penetration if the welded joint is either lightly loaded or is just a sealing weld.

The main purpose of the joint preparation is to permit the required level of fusion between the joint faces. The type of weld preparation applied will therefore be dependent upon the thickness of the material and the welding process to be used. Some typical butt weld joint preparations are shown in Fig. 1.3.

Single-sided preparations allow welding to take place from one side whereas double-sided preparations require welding to be done from both sides. A major disadvantage of having access to only one side is that the other side of the initial root run cannot be accessed to carry out removal of welding-induced defects. An example of this would be where a large gap was required to permit fusion throughout the whole cross-section of the weld but excessive penetration then occurred. Good control and formation of the root weld must therefore be maintained at the time of welding, and this can be assisted by using the root control measures shown in Fig. 1.4.

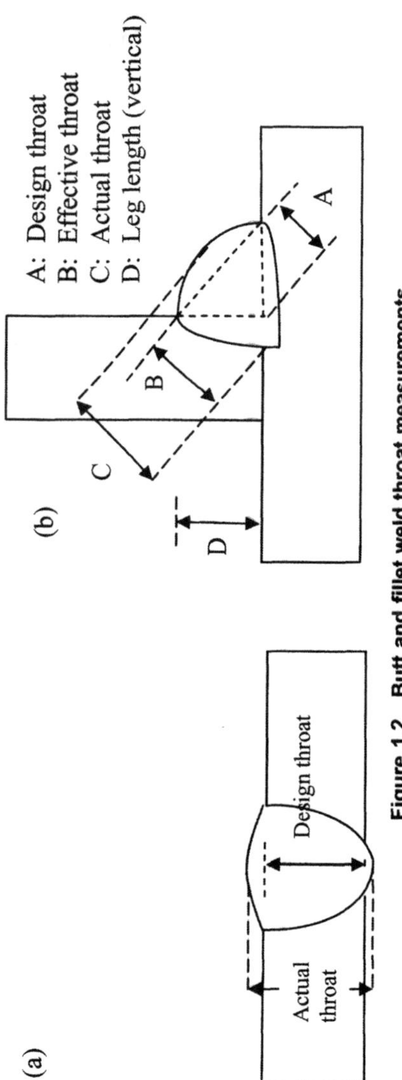

Figure 1.2 Butt and fillet weld throat measurements

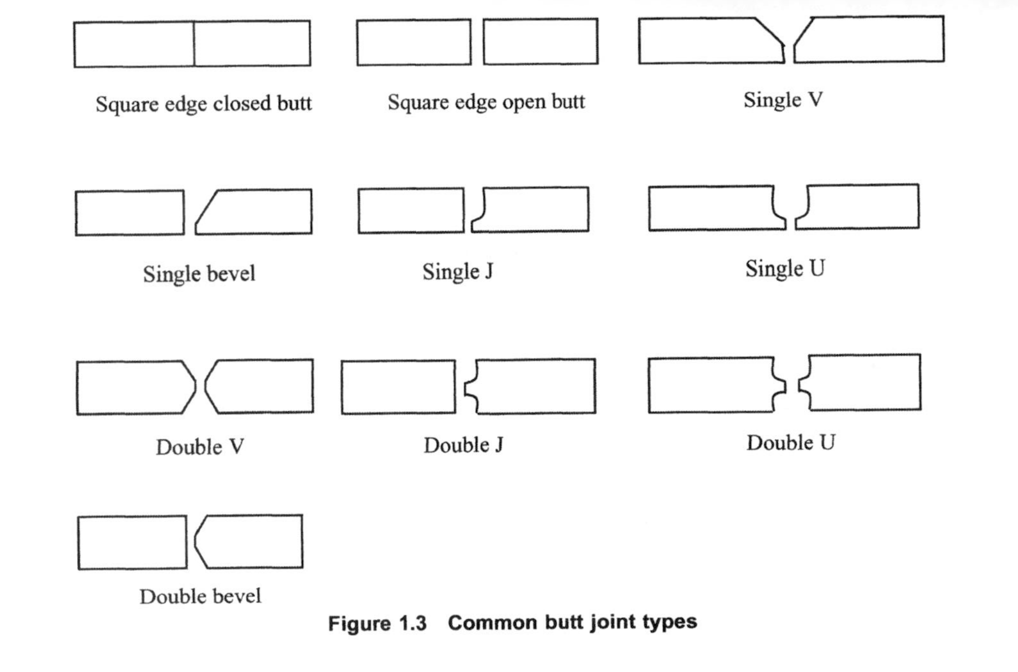

Figure 1.3 Common butt joint types

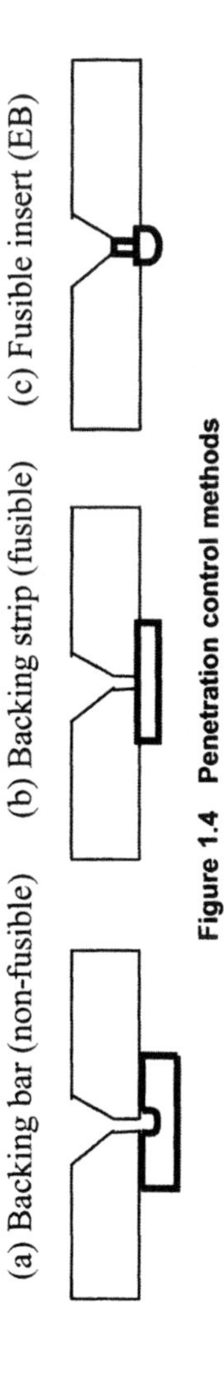

Figure 1.4 Penetration control methods

- Non-fusible backing bars. These are always removed and are normally made of copper (often water cooled) or ceramic and help form the root weld shape. It is important to ensure that the copper bar does not melt and contaminate the weld as this could cause weld metal cracking.
- Fusible backing strips. These are made of a material similar to the parent material and are tacked on to the parent material. They are fused into the root and are normally ground off, or occasionally left in place.
- Pre-placed filler such as EB inserts. These are used on pipe and welded using a TIG (tungsten inert gas) process. The EB stands for Electric Boat and is the name of the company that first supplied them. Use of EB inserts is a specialised procedure used in specialist applications such as the nuclear industry.

Fillet weld shapes are either mitre, concave or convex, as indicated in Figure 1.5.

- Mitre fillets are the most common and normally have equal leg lengths. They give a good combination of maximum design strength and toe blending for fatigue resistance.
- Concave fillets have a reduced throat measurement for their leg size, which gives them a reduced strength. The smoothly blended toe design gives them very good fatigue failure resistance.

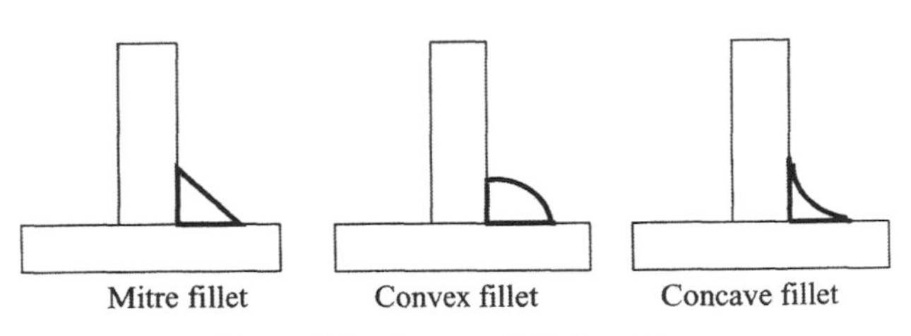

Figure 1.5 Types of fillet weld

- Convex fillet welds have sharp toes, which give very poor fatigue resistance but the excess weld metal they have on the throat gives maximum design strength (although it will increase the weight of the weld). The excess weld metal will also increase cost due to the extra consumables required and increased welding time. These welds are only used where strength is paramount and fatigue loading is not an issue.

Drawing rules and weld symbols

Weld symbols on drawings are a very efficient way to transfer fabrication information from the designer to the fabricator by showing the joint and welding information as a symbolic representation. This reduces the amount of information that would have to be put on the drawing in written form or as sketches. An inspector must have a good working knowledge of weld symbols as a large proportion of inspection time is spent verifying that the welder is complying with the approved fabrication drawing. Weld symbols themselves are similar between the major standards but there are some differences in how they are presented. It is important to understand the basic differences between the major standards and to be able to recognise any drawing standard being used. Reference should always be made to the applicable standard for specific symbolic information.

Common standards in use today are BS EN 22553 (which replaced BS 499) and AWS A2.4. Reference is still made to BS 499 because many old fabrication drawings will have been done to this standard. Most drawing standards follow a basic set of rules or conventions when formulating welding symbols. A weld symbol is composed of five main components common to most standards, consisting of an arrow line, a reference line, the welded joint symbol, dimensional information and finally any supplementary information.

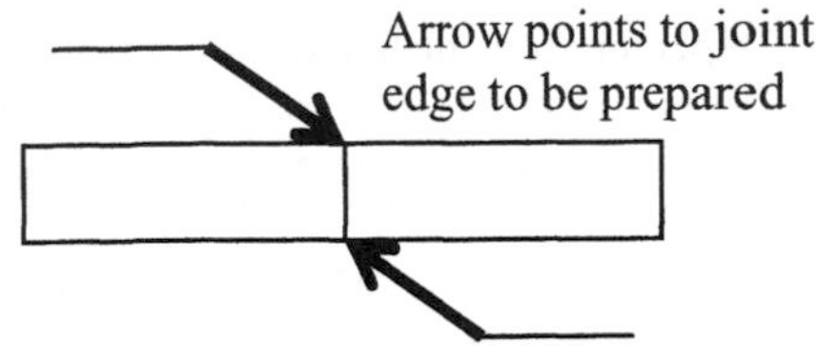

Figure 1.6 The arrow line

The arrow line

The arrow line is a single, straight, solid line (there is an exception to this in AWS A2.4 for single plate preparations where the arrow line is staggered) (Fig. 1.6). It touches the joint intersection and points to the plate edge that is to be prepared. For symmetrical joints it can point to either plate edge as they are both prepared in the same way. It must not be parallel to the bottom of the drawing and must always be finished with an arrow head.

The reference line

The reference line is a solid line that touches the arrow line (Fig. 1.7). It is preferably drawn parallel to the bottom of the drawing, but if this is not possible it is drawn perpendicular to the bottom. There must be an angle between the arrow line and reference line where they join. In BS EN 22553 a dashed line is also placed either above or beneath the solid reference line and relates specifically to the 'other side' of a joint.

The welded joint symbol

The joint symbol is used to represent the different joint categories and is generally similar in shape to the weld that it

Figure 1.7 The reference line

represents. Depending on the standard used, it is normally positioned below or above the reference line to specify whether the weld is on the 'arrow side' or 'other side' respectively. In BS 499 and AWS A2.4 the symbol is placed below the reference line to represent the fact that the weld is made from the arrow side (i.e. the side that the arrow is touching). If the symbol is on top of the line it indicates that the weld will be made from the opposite side (i.e. the other side) to which the arrow is pointing (or touching). This often causes confusion and can lead to welds being put in the wrong position.

To avoid this confusion the 'other side' is represented by a separate dashed line in BSEN 22553 (see Fig. 1.8). It therefore does not matter if the symbol is above or below the solid line because it will always be on the arrow side except when placed on the dashed line. A symmetrical joint such as a double V, which is the exact same size on both sides, does not require a dashed line to represent the other side because both sides are the same.

Supplementary symbols may be used to complete a joint symbol with additional information such as a flat, concave or convex finish (Fig. 1.9). The absence of a supplementary symbol indicates that the weld is left in the 'as-welded' condition with no precise weld surface finish required.

Dimensions

Dimensions that relate to the joint cross-section, such as weld depth or fillet sizing, are positioned on the left of the symbol. If no weld depth is specified before the symbol on a butt weld it is assumed to be a full penetration weld. Fillet welds will always have a dimension relating to the leg length and/or throat dimension. Longitudinal dimensions such as weld lengths are positioned on the right of the symbols in most standards (see Fig. 1.10). If no longitudinal dimensions are specified after the symbol then the weld extends the full length of the joint.

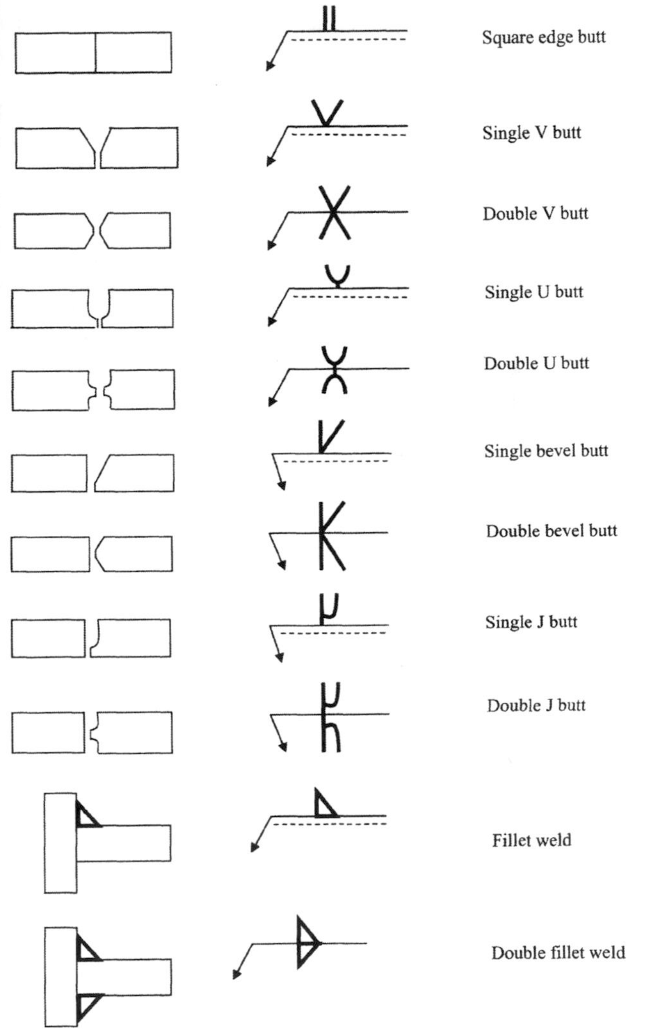

Figure 1.8 Typical weld drawing symbols

Supplementary Symbols (from BS EN 22553)		
	Shape of weld surface or weld	Symbol
(a)	Flat (usually finished flush)	—
(b)	Convex	⌒
(c)	Concave	‿
(d)	Toes shall be blended smoothly	⅃
(e)	Permanent backing strip used	M
(f)	Removable backing strip used	MR

(a)

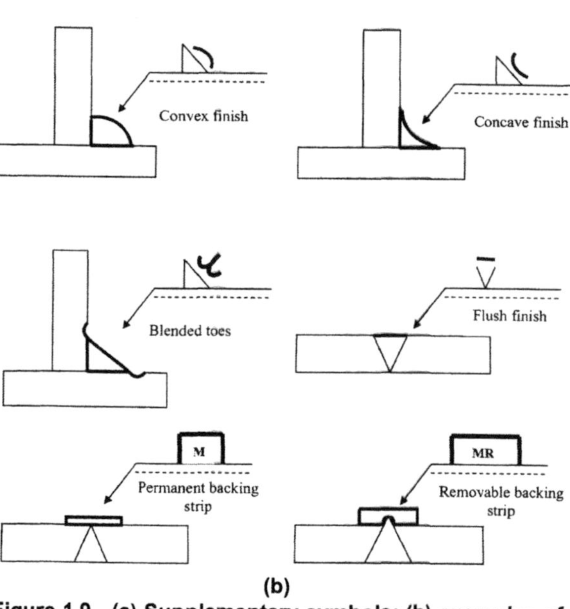

(b)

Figure 1.9 (a) Supplementary symbols; (b) examples of supplementary symbol use

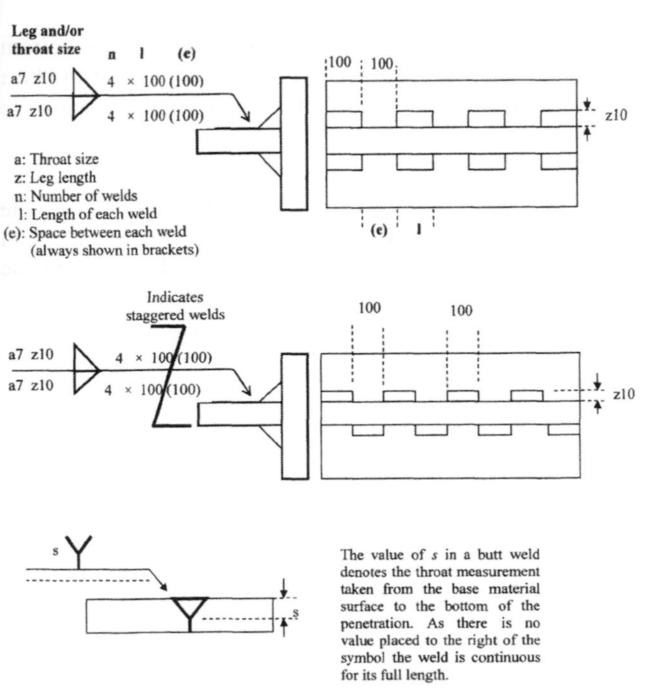

Figure 1.10 Weld dimensioning examples (BSEN 22553)

Supplementary information

Supplementary information specifying NDT, welding processes or special instructions can differ between standards and may be shown at the end of the reference line or adjacent to it (see Fig. 1.11). BSEN 22553 refers to this as complementary information. A site or field weld is indicated by a flag positioned at the joint between the arrow and reference lines. A circle at the same position is used to indicate that a peripheral weld is required all around a part.

Other common rules

When a symbol has a perpendicular line (single bevel, single J, fillet weld, etc.) it will always have the line positioned to the left. When the same symbol is shown on the top and bottom of the reference line (double V, double bevel, etc.) it will be a mirror image of itself (based on the reference line being the mirror).

BSEN 22553 specifics (Fig. 1.12)

- BSEN 22553 does not have a single reference line with the symbols placed above and/or below it to specify if the weld is on the 'arrow side' or 'other side', but adds a dashed line

Complementary information can be added to a tail at the end of the reference line. This information could include (in order of priority and separated by a solid line)

- Welding process
- Acceptance level
- Working position
- Filler materials

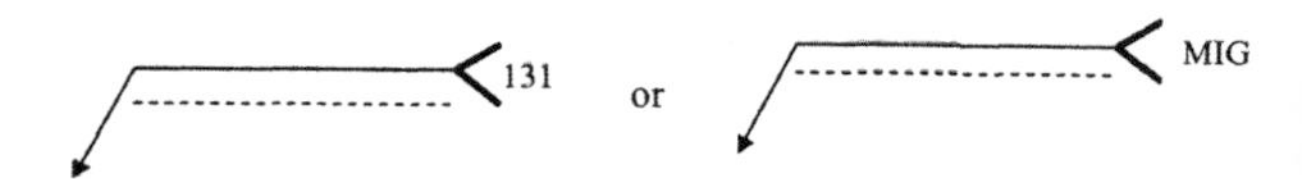

A closed tail at the end of the reference line is also possible and indicates a specific instruction such as the use of a particular procedure.

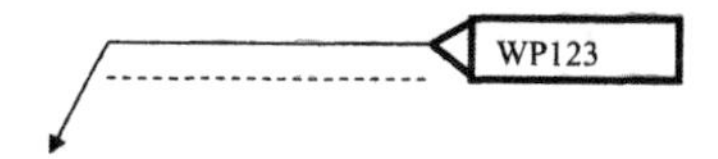

Site welds and peripheral welds are indicated on the junction of the arrow line and reference line

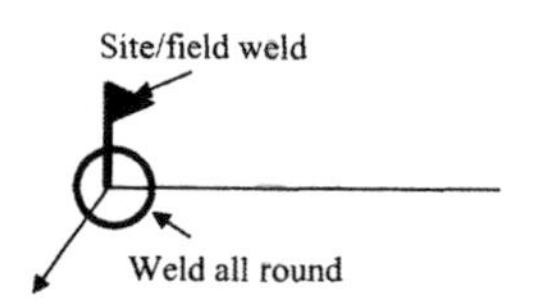

Figure 1.11 Supplementary information

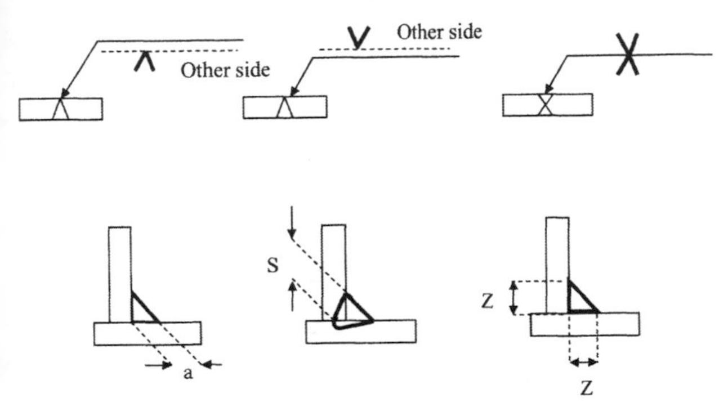

Figure 1.12 Specifics of BSEN 22553

to specify the 'other side'. This makes good sense because engineering drawings represent what you cannot see as a dashed line and obviously you cannot see the 'other side' of a joint. It helps reduce the risk of drawings being read wrongly and welds being incorrectly placed on the wrong side of a joint.

- The dashed line specifying 'other side' can be omitted if a double-sided weld is symmetrical about the reference line.
- Fillets leg sizes must be prefaced with the letter *z*.
- Nominal fillet throats are preceded by the letter *a*.
- Effective throat thickness for deep penetration fillet welds and partial penetration butt welds is preceded by the letter *s*.

AWS A2.4 specifics (Fig. 1.13)

- AWS A2.4 may use more than one reference line from the arrow line to indicate the sequence of welding.
- Weld dimensions may be given as fractions or decimals and in metric or imperial units.
- Welding processes are indicated using standard AWS abbreviations.
- Single plate preparations are indicated by a directional

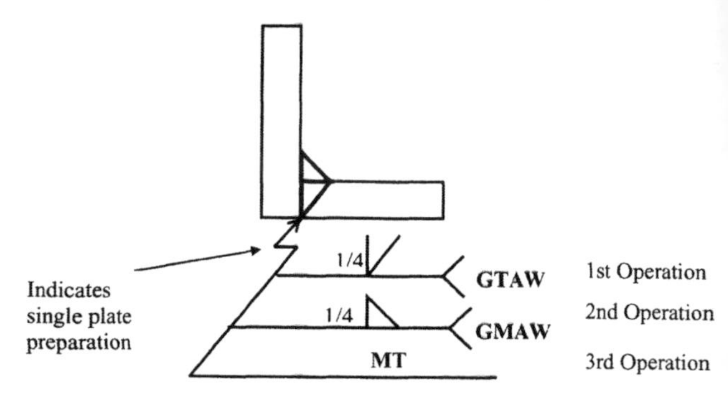

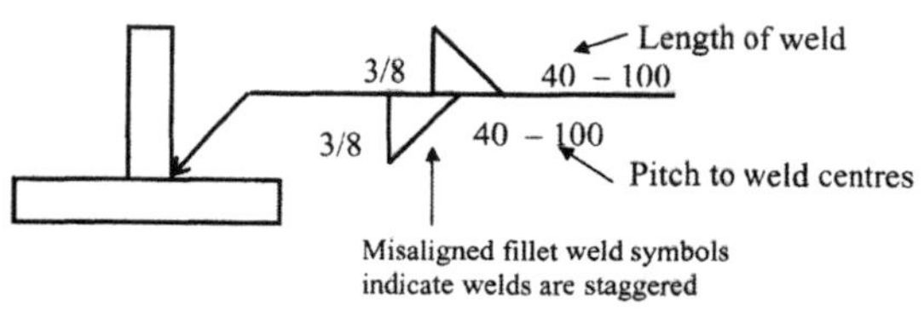

Figure 1.13 Specifics of AWS A2.4

change of arrow line but the arrow remains pointing to the plate edge requiring preparation.

- When plate preparation within a joint is obvious (i.e. a T joint) then the direction of the arrow line is optional.
- AWS A2.4 dimensions the pitch of intermittent fillet welds and plug welds to the centre of each weld. (The BS and BS EN codes dimension these to the start of each weld.)

Chapter 2

Duties of a Welding Inspector

A welding inspector has the responsibility to monitor all aspects of the welding process before, during and after welding to ensure that the finished weldment is fit for purpose. Fitness for purpose in this sense normally means that the finished weldment has been produced in accordance with a code or specification and complies with the stated requirements of that code or specification. The inspector must therefore be proficient in carrying out visual inspections and assessing his or her findings in accordance with the relevant code or specification acceptance criteria. The following checks are commonplace.

Before welding

Before welding is carried out the inspector checks that all of the welding variables specified on the weld procedure specification (WPS) can be achieved and that the correct equipment and documentation is available. The level of checking will vary depending on the code requirements but may involve the following checks.

Documentation

- Drawings, codes, specifications and standards are available and at the correct revision state.
- Correctly qualified welding procedures are in place covering the welding processes to be used in production.
- Welders are correctly qualified for the welding processes and the WPSs to be used in production.
- Correct NDE procedures and techniques are in place.
- NDE and heat treatment personnel are suitably qualified.
- Correct preheat and post-weld heat treatment procedures specifying temperatures, holding times and temperature measurement methods.

Equipment

- Welding machines are in date for calibration or validation (welding machines with meters are normally calibrated while machines without meters are validated).
- All ancillary equipment such as ovens, quivers, remote controls, torches, cables, etc., are in good condition.
- Any heat treatment equipment is in good condition and in date for calibration if required.

Materials

- Verify the correct materials are being used and are in an acceptable condition.
- Ensure the material certificates are the correct grade and the batch or heat numbers correspond to the actual material.
- Verify the correct consumables (covered electrodes, wires, fluxes or gases) are being used and have been correctly prepared.
- Verify any consumables certificates match consumables batch numbers.

Safety

- Ensure all safety precautions are being adhered to and that any work permits required are in place.
- Ensure all electrical equipment is in date for test.
- Ensure correct and sufficient ventilation is in place.
- Ensure the correct PPE is being used.

Weld joint preparation

- Ensure the correct weld joint preparation has been formed in accordance with the specified procedure using an approved method (flame cut, ground, machined, etc.).
- Ensure the joint is correctly sized (groove angles, groove radius, root face, etc.) and the fit-up is within tolerance (root gap, alignment, etc.) using correct restraints.
- Ensure any required pre-heat is applied using the correct

heating method and temperature measurement methods are approved for use.

- Ensure temperatures are in accordance with the WPS.
- Ensure any tack welds are correctly applied after any pre-heat requirement.

During welding

The inspector must monitor all aspects of the actual welding to check that the correct heat input is applied to the weld and that the weld is formed correctly in accordance with the WPS. Realistically, it is pointless making all the necessary preparations and checks before welding and then simply assuming that the welder will strictly adhere to the procedure. Many a weld has failed because a welder has taken a shortcut without appreciating the detrimental effect to the mechanical properties of a welded joint. One of the functions of the inspector is therefore to police all the parties involved in the formation and testing of the weld. Checks to be carried out include:

- Monitor and record the welding variables such as amperage, run-out lengths and voltage.
- Monitor and record environmental conditions (rain, snow, wind, etc.) that could affect the finished weld.
- Monitor and maintain pre-heat temperatures.
- Monitor the interpass temperatures and maintain them as required.
- Ensure interpass cleaning is carried out correctly.
- Inspect critical root runs and monitor hot pass times.

After welding

Once all welding is completed it is the responsibility of the inspector to ensure that the required quality inspections and tests are carried out by qualified personnel and meet the relevant code or specification acceptance requirements. The main checks are to:

- Ensure any required PWHT is correctly applied and

maintained by suitably qualified persons using suitable procedures and equipment. PWHT may be required before and after NDE when hydrogen cracking is a possibility.

- Ensure required NDE is carried out and review the results.
- Visually inspect and measure the completed weld and assess any imperfections in accordance with the relevant code acceptance criteria.
- Maintain weld records with weld identification.
- Ensure any required pressure testing of piping and pressure vessels is carried out after PWHT and results recorded.
- Take final dimensional checks after pressure testing.
- Collect and collate all the relevant documentation and pass the completed package to the relevant acceptance authority.

Repairs

When a defect is identified during the inspection, three main options exist, depending upon the severity of the defect and the component being welded. The component can be scrapped, it can be given a concession or it can be repaired.

- A **concession** is a situation where a defect is deemed not to have too detrimental an effect on the component and the client agrees in writing that the defect can remain in place. In some cases the concession conditions may mean a limitation is placed on the component service conditions and/or a repair is required at a more convenient time.
- A **repair** can in some cases require a different welding process, different consumables or some other change to the original procedure to be used. There may be limitations on the number of repairs or thermal cycles applied to the component depending on the material used. A repair procedure should be in place detailing what actions are required.

The simplest way to think about carrying out a repair is that the inspector monitors all aspects of the repair process before, during and after welding the component. This gives the best possible assurance that the weld is fit for its specified purpose.

Chapter 3

Analysis of a Fusion Weld

Components of a welded joint

A welded joint comprises the weld, the heat affected zone (HAZ) and the base metal. The components of the weld and HAZ are shown in Fig. 3.1. The main features are as follows.

Excess weld metal

This is excess to requirements and does not add to the strength of the weld. It is also known as the cap (a slang term), reinforcement (BS term) or crown (US term). Reinforcement is a poor term because it implies that adding reinforcement strengthens the weld, which is not the case. In practice, the addition of reinforcement (or excess weld metal) can effectively weaken the weld by increasing the stress concentration at the weld toes, hence reducing the fatigue life of the joint.

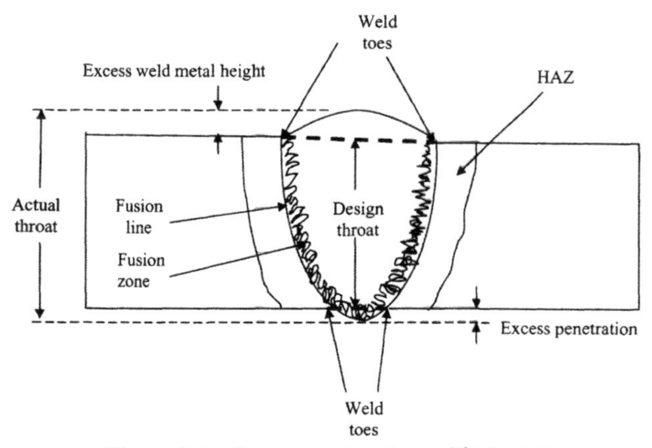

Figure 3.1 Components of a welded joint

Weld toes

These are the points where the weld metal adjoins the base metal. There will normally be four weld toes on a full penetration butt weld (two on the face and two on the root) and two on a fillet weld. The greater the angle at the weld toe, the greater the risk of a fatigue failure if the component is subjected to cyclic stress.

Fusion line

This is where fusion takes place between the melted and unmelted material. It is commonly referred to as the fusion boundary or weld junction.

Fusion zone

The fusion zone is the region within the weld that contains the greatest dilution of filler metal with melted base metal. This region of highest dilution can contain defects owing to impurities or contaminants contained in the base metal being drawn into the weld. The centre of the weld will be the area of the lowest dilution and may consist of filler metal only.

HAZ

This is the region of the base metal that has not been melted but has been affected by the heat and had a change made to its grain structure. Cracks often occur in the HAZ after welding has taken place owing to it being hardened by the formation of a martensitic grain structure on cooling.

Excess penetration

Excess penetration is excess weld metal formed in the root of the weld. Some codes may specify a limit to the penetration due to the toe blending and/or the bore restriction caused in piping.

Design throat thickness

The throat is where the strength is contained within the weld. It is equal in size to the thinnest of the base materials being joined.

Actual throat thickness

This is the actual measurement made from the weld face to the root. The actual strength of the weld can be calculated using this actual throat measurement (but any measurement in excess of the design throat is ignored). This is more relevant for partial penetration welds or where root concavity is present.

What makes a good fusion weld?

There are four factors that must be satisfied to produce a good fusion weld. If any of these factors is not achieved then the result will be a weld that may not be fit for purpose. The four factors are as follows:

- Heat input. This is the heat that melts the parent material and filler (if required) to give the required fusion between the parts being joined. The heating could be from an arc, a laser, an electron beam or an oxyacetylene gas mixture. The most common process is arc welding.
- Protection from atmosphere. If exposed, a welding arc will pick up gases such as oxygen, nitrogen or hydrogen from the atmosphere and pass them into the weld pool. These gases can have a detrimental effect on the finished weld so the arc needs to be protected from them. Common methods of atmospheric protection used in arc welding processes include the use of:
    - a shielding gas for processes such as TIG, MIG, MAG and PAW;
    - a gaseous shroud, which is produced as the flux coating on the electrode melts in MMA or FCAW;
    - a flux blanket, which covers the arc in SAW;
    - a combination of a shielding gas and a gaseous shroud from melting flux in secondary shielded FCAW.
- Protection from external and internal contaminants (cleaning). The weld metal can pick up contaminants from the surface of the material so it is important that the material surface is free from scale, rust, paint, grease,

moisture and other possible weld contaminants. Internal contaminants such as sulphur or oxides can be present within the material and may also have to be removed or neutralised during welding. In effect, the process must be able to 'clean' the material and weld pool during welding to afford the correct protection. Protection from contaminants can be achieved by:

- mechanical cleaning of the component (grinding, wire brushing, abrasion, etc.);
- chemical cleaning of the component and filler wire/rods (acid, acetone, etc.);
- use of a flux (containing deoxidisers);
- use of correct polarity (d.c. +ve or a.c. when welding Al or Mg alloys to remove the surface oxide layer, called cathodic cleaning).

- Adequate mechanical properties. The finished welded joint must have adequate properties such as strength, toughness, hardness and ductility in the base material (including the HAZ) and weld metal. These properties are achieved (depending on the welding process) by:
  - using the correct base materials;
  - using the correct consumables (filler wire, electrodes, shielding gas, flux);
  - using correctly prepared consumables (correctly heat-treated electrodes, etc.);
  - using the correct pre-heat and/or post-weld heat treatments;
  - using the correct heat inputs (voltage, current and travel speeds).

All of the above factors can be achieved by adhering strictly to the requirements of an approved welding procedure specification, which contains the essential, supplementary essential (ASME IX only) and non-essential variables necessary to produce a sound weld.

In practice, when welders get lazy, take shortcuts and do not fully comply with a correctly tested and qualified welding procedure, problems can arise. A common example is when tack welds are done without applying the specified pre-heat. This leads to the tacks and HAZ around them being harder than they should be because they cool down too quickly (remember, the main reason for pre-heat is to retard the cooling rate and reduce temperature variations across the weldment). These hardened areas within a highly stressed region can then suffer cracking, with possible catastrophic consequences if they are not found.

Weld joint: preparation methods

The type of joint preparation and how it is made can have an effect on the final weldment properties. For example:

- Weld preps made using thermal cutting processes can be affected by the heat of the cutting process. This can lead to a possible loss of toughness or cracking in the HAZ, so approximately 3 mm of material is normally removed from the prep by mechanical means after thermal cutting. Typical thermal cutting processes (used to form straight-sided preps such as square edged or bevelled) include plasma, oxy-fuel gas or oxy-arc processes.
- If cutting is by an oxygen fuel gas process such as oxyacetylene then the weld preps can gain excess carbon from the process, leading to cracking in the weld (the carbon increases the weld metal hardness and therefore the risk of cracking). This is another reason to remove 3 mm from the prep by mechanical means after cutting.

  Remember that oxy-fuel gas cutting does not actually melt and blow away the molten material like plasma or oxy arc cutting does. What it actually does is to heat up the material to its ignition temperature and then introduces a stream of oxygen, turning the metal into instant rust, which is then blown away. This is why stainless steel (a rust-resistant material) cannot be cut by oxy-fuel gas unless special powders are added to the process.

- If mechanical cutting such as machining (normally required for weld joint preparations containing a radius such as a U or J prep) or shearing is used then consideration must be given to the possibility of having cutting fluids trapped in the cut edges. These can cause porosity or other weld defects unless they are properly cleaned out (the edge may look smooth but under a microscope it will be full of peaks and troughs).

Weld joint: shape

The shape of the joint (joint type) can affect not only the final mechanical properties of the weldment but also the level of distortion produced in it. The mechanical properties can be affected because a change of joint type may affect the overall heat input. This is due to the possible requirement for more (or less) weld metal to be welded into the joint. The actual level of final distortion is determined by factors such as the weld metal shrinkage direction (based on the angle of the groove) and whether the weld is formed from one or both sides.

Figure 3.2 shows the components of a typical single V joint and U joint, and some advantages of the single U over a single V in thicker materials.

Residual stress and distortion

Residual stresses are those stresses that remain in a component after a procedure such as bending or welding has been carried out. If a plate is bent then bending stresses will be introduced into the bend and the maximum tensile stress will be at the outer convex surface. This stress remains in the plate and can cause failure, especially if a defect or stress raiser is present in the stressed region (this of course is how bend tests are carried out).

Residual welding stress in a welded joint comes from the uneven expansion and contraction of the joint coupled with a restraint. Consider the following example.

An unrestrained strip of metal is heated up and expands. If

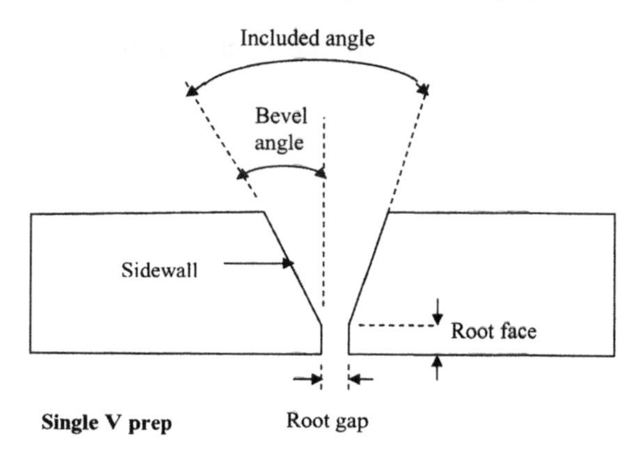

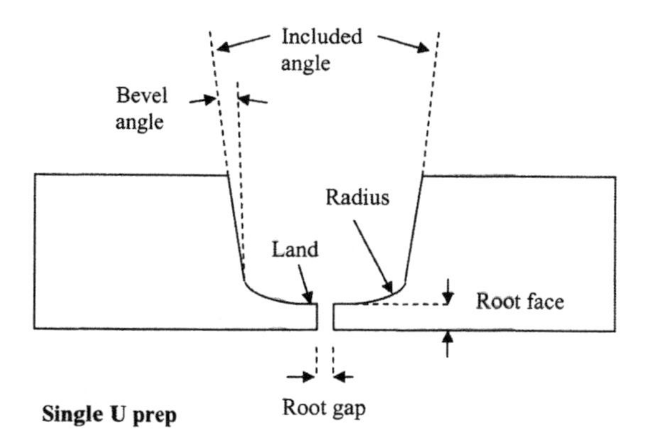

Some benefits of single U versus single V in thicker materials

- Less volume to fill
- Less filler required
- Lower heat input
- Less distortion

Figure 3.2 Single V and U preparations

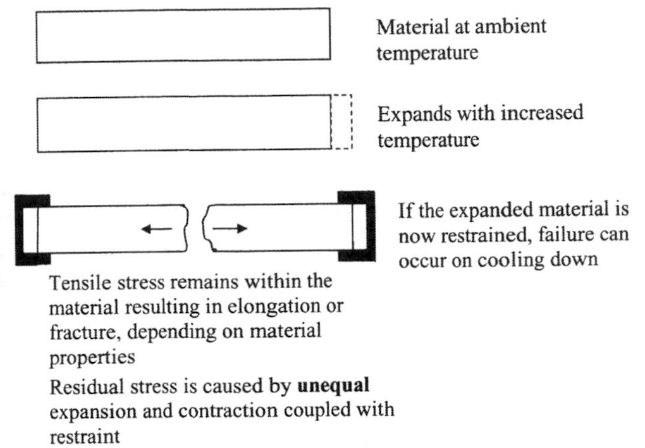

Figure 3.3 Unequal expansion and contraction

it is then allowed to cool it will shrink back to its original size and there will be no residual stress present. If, however, the expanded metal was to be restrained by jigs (see Fig. 3.3) to try and prevent it shrinking when it cools, then a tensile stress will be introduced into the material and can cause the material to deform plastically or fracture. This, in effect, is residual stress induced through uneven expansion and contraction coupled with restraint.

Now consider the mechanics of welding where a pool of molten weld metal cools down and contracts (shrinks). The base material acts as a restraint and tries to prevent the weld pool contracting, providing the ideal conditions required to leave residual stress in the completed weldment. It can be deduced from this that there will always be residual stresses present in welded joints. In some cases this may be high enough to approach the yield point.

Mechanical restraints such as jigs and clamps increase restraint stresses further by preventing the movement that would normally occur as the molten weld metal cools down and shrinks. This causes the joined components to distort. Allowing distortion to take place helps to reduce the overall stress in the component, but may make the final dimensional criteria unacceptable with the code requirements. Methods can be employed that can reduce both residual stress and distortion (see later).

Residual stress in a weld acts in many directions in a complex pattern because any weld will change volume and shape in all directions. There are three main stress directions caused through weld metal shrinkage to consider: the longitudinal, transverse and short transverse directions (see Fig. 3.4).

Distortion

If an unrestrained material is heated and cooled uniformly then there will be no distortion produced within it. If, however, the material is subjected to localised heating and cooling then distortion will occur due to the different rates of expansion and contraction experienced throughout the material, caused by temperature gradients. Welding does not heat and cool the material uniformly because the weld region will always be hotter than the surrounding region and the weld metal will therefore expand and contract at a much greater rate than the adjoining base material. The base material can be pre-heated to reduce the temperature gradient spreading from the weld outwards but obviously

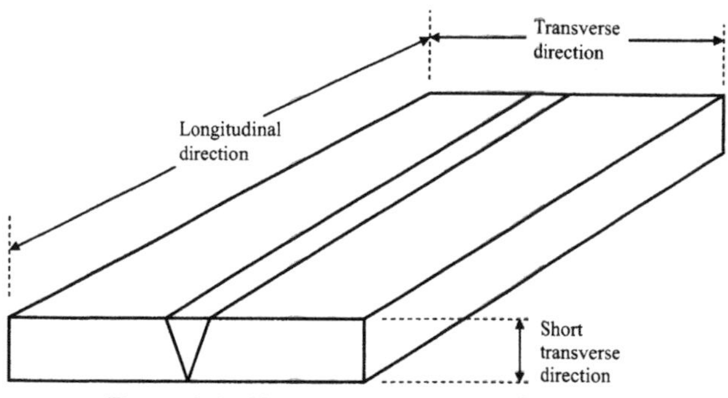

Figure 3.4 Main stress directions in a weld

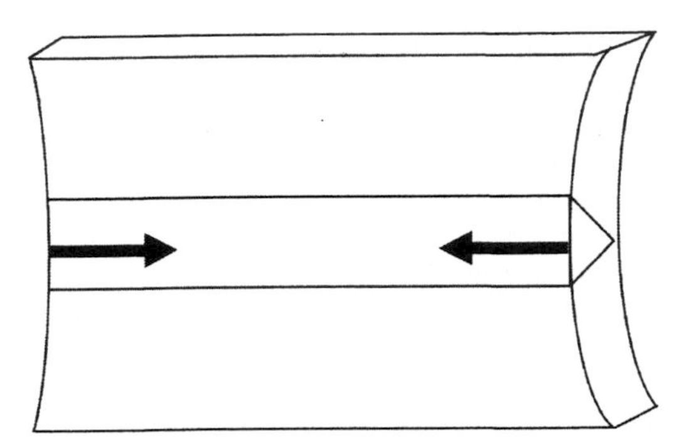

Longitudinal distortion

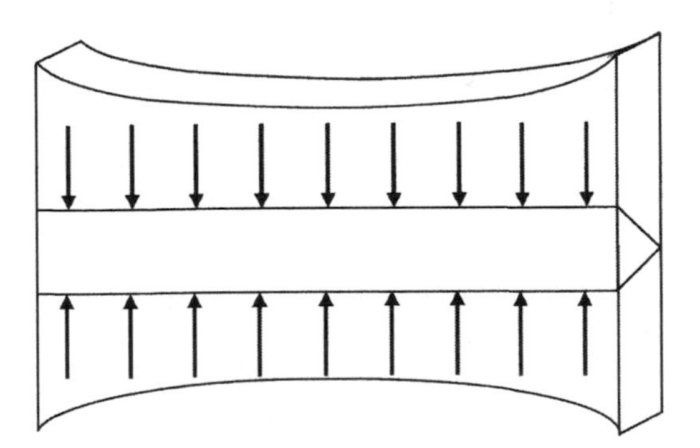

Transverse distortion

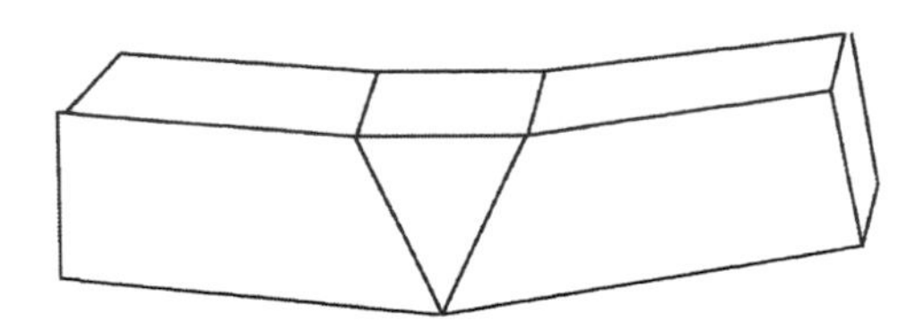

Angular distortion

Figure 3.5 Distortion caused by the effects of shrinkage

cannot be uniformly heated to the melting temperature required to weld. The net effect of this is that some distortion will always occur because the weld effectively acts as a form of localised heating.

The coefficient of thermal expansion of a material plays a large part in how much welding stress is introduced into a material and how much distortion can occur. The higher the coefficient of expansion, the higher the distortion level, which is why stainless steels suffer higher distortion levels than plain carbon steels.

Simplistically, the main distortions to consider are caused by the weld metal shrinkage in the longitudinal, transverse and short transverse directions. In reality, the shrinkage and distortion will follow a very complex pattern but a simplified exaggerated view of the effects is shown in Fig. 3.5.

The shape of a weld preparation will affect how much distortion is produced, because bevel angles will help to direct the distortion direction. The material thickness, amount of weld metal required and size of individual weld runs also affect stress and distortion levels. A single V butt will suffer considerable distortion, especially during the initial root runs when there is no restraint, because of the high level of weld metal shrinkage. Although further runs add to the distortion, the overall effect is not cumulative because the previous runs will help to restrain the joint from moving. A square-edged closed butt will actually suffer very little distortion. Figure 3.6 compares distortion levels based on the bevel angle of two single V butt welds and a square-edged closed butt weld.

Minimising stresses and distortion

There are various ways to counteract the effects of residual stress levels and distortion depending upon the type of welding process and joint type that is being used.

Offsetting (Fig. 3.7) is where plates are offset to a preset angle and welded without restraint. As the weld solidifies and shrinks it distorts the plates and pulls them into the correct

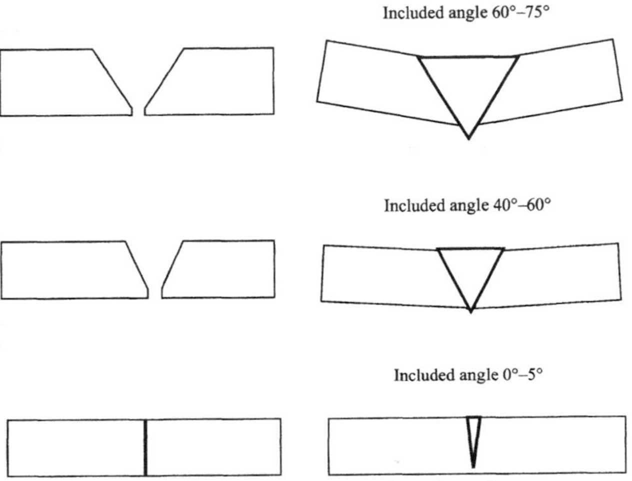

Figure 3.6 Distortion and bevel angles

position. It can be used on fillet welded T-joints and butt welds but the amount of offset required by welding is normally only determinable by trial and error. It is a very cheap and simple way to control distortion if carried out

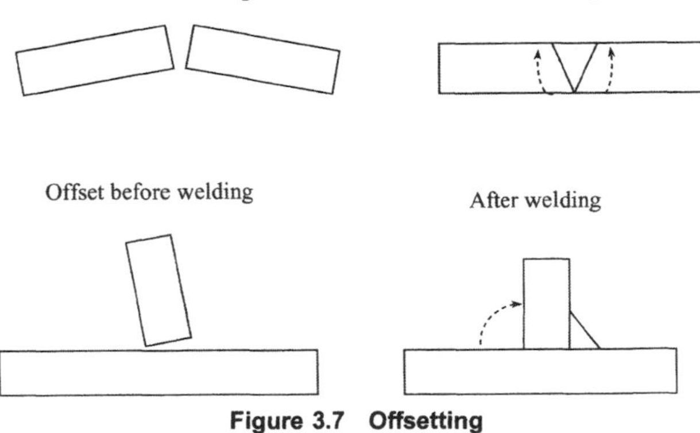

Figure 3.7 Offsetting

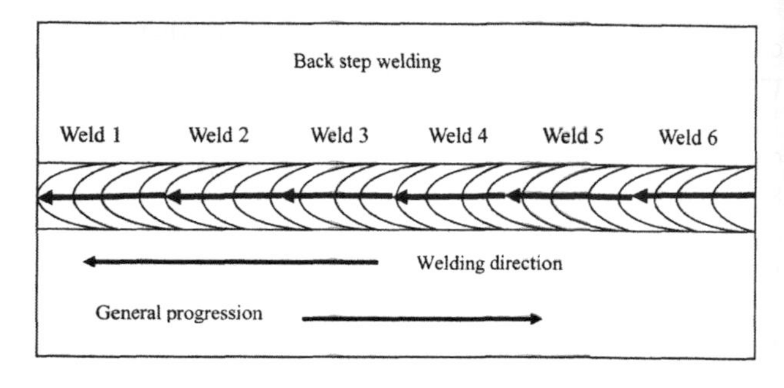

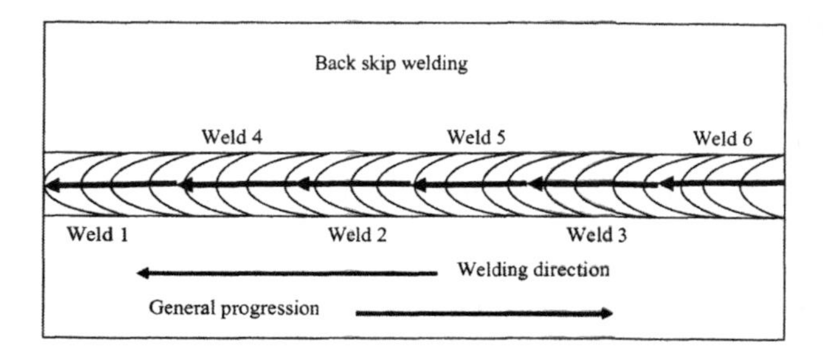

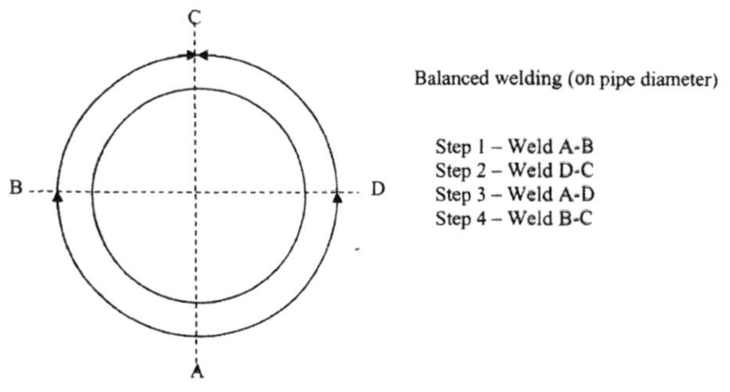

Figure 3.8 Sequential welding examples

correctly. By allowing the plates to move and not restraining them the level of residual stress in the welded joint is reduced.

Sequential welding (Fig. 3.8) is a technique utilised to control the level of distortion by welding the joint in a specific way. There are various sequential welding techniques including balanced welding (welding about a neutral axis of the weld), back skip welding or back step welding (a short length is welded and the next weld starts a short distance behind the previous weld length).

Mechanical restraints including clamps and jigs allow accurate positioning of the component parts. They are normally left in position until the components are tack-welded together and then removed to allow the welder full access. Restraints will reduce the level of distortion by physically holding the components in position but will increase the stress levels introduced into the weld.

Flame straightening uses an oxyacetylene torch (but not a cutting torch) to give a localised intense heat. The distortion caused by this localised heating can be used in some cases to straighten or modify the shape of a component. Typical examples of this method would be straightening flanges or removing bulges from insert plates.

Heat treatment methods such as post-weld heat treatment stress relief can remove a high percentage of residual stresses.

Mechanical stress relieving methods include:

- using ultrasound to stress relieve fabrications;
- peening weld faces using pneumatic needle guns, to redistribute residual stresses by placing the weld face in compression.

Chapter 4

Materials and Their Weldability

There are many different definitions concerning the weldability of steel because it often means different things to different people. Simplistically it can be defined as the ability of a material to be welded and still retain its specified properties. This ability to be welded successfully depends on many factors including the type and composition of the material, the welding process used and the mechanical properties required. Poor weldability generally involves some type of cracking problem and this is dependent upon factors such as:

- residual stress level (from unequal expansion and contraction due to welding);
- restraint stress level (from local restraint such as clamps, jigs or fixtures);
- presence of a microstructure susceptible to cracking (the base material may have a susceptible microstructure or the HAZ and/or weld metal may form a microstructure susceptible to cracking owing to the welding).

Carbon equivalency

The susceptibility of a microstructure to cracking is heavily influenced by the amount of carbon and the type and amount of other alloying elements present in the steel. The carbon and other alloying elements can be put into a formula that determines the carbon equivalency value (Cev) of the material. This Cev is a measure of the hardenability of the steel. The higher the Cev, the more susceptible the material will be to cracking by brittle fracture.

Other factors affecting the likelihood of cracking include the base metal thickness and the combined joint thickness (i.e. a butt weld has two thicknesses whereas a fillet weld has three). The combined joint thickness is important because

$$Cev = C + \frac{Mn}{6} + \frac{Cr + Mo + V}{5} + \frac{Ni + Cu}{15}$$

This is not the full formula but contains as much as a welding inspector will need to use.

Figure 4.1 Simplified carbon equivalency formula

each material thickness acts as a heat sink and can cool the weld area more quickly, making it harder and therefore more susceptible to cracking.

Figure 4.1 shows the formula used to determine the Cev for a material. As a general guide the following Cev levels determine the weldability of steels:

- up to 0.4%: good weldability;
- 0.4 to 0.5%: limited weldability;
- above 0.5%: poor weldability.

Classification of steels

Low carbon steel: contains 0.01–0.3% carbon
Medium carbon steel: contains 0.3–0.6% carbon
High carbon steel: contains 0.6–1.4% carbon

Plain carbon steels contain only iron and carbon as main alloying elements. Traces of other elements such as Mn, Si, Al, S and P may also be present. It can be seen from the carbon diagram in Fig. 4.2 that an increase in carbon causes the ductility of steel to decrease while the tensile strength and hardness increase. Note also how the maximum tensile strength of plain carbon steel is achieved at 0.83% carbon content.

Alloy steels contain alloying elements such as Mn, Mo, Cr and Ni and are divided into two groups:

- Low alloy steels contain $<$ 7% total of other alloying elements.
- High alloy steels contain $>$ 7% total of other alloying elements.

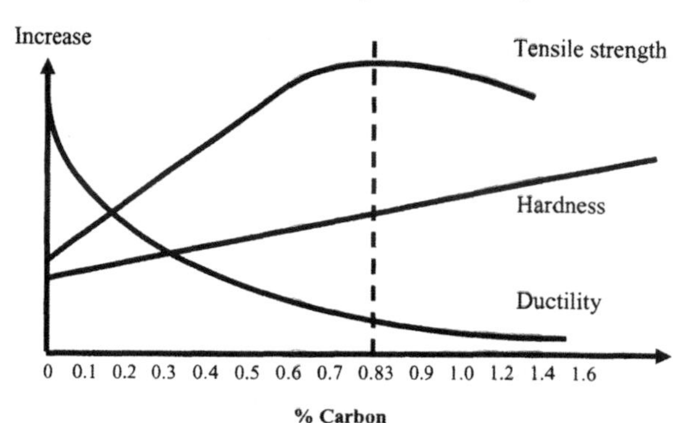

Figure 4.2 Carbon diagram

Alloying elements

The following are some basic properties of alloying elements added to steels:

Iron (Fe)	This is the basic constituent of steels.
Carbon (C)	Increases tensile strength and hardness but reduces ductility.
Manganese (Mn)	Improves toughness and strength when alloyed at levels of < 1.6% in steels. Can control solidification cracking in steels by neutralising the detrimental effects of sulphur.
Chromium (Cr)	Alloyed at levels > 12% to produce stainless steels. Gives corrosion resistance and promotes through-thickness hardenability. Hardenability is the ability of steel to harden at slower cooling rates when alloying elements are added to it. Do not confuse this term with hardness.
Molybdenum (Mo)	Gives high temperature creep resistance and strength in low alloy steels.
Nickel (Ni)	Improves strength, toughness, ductility

	and corrosion resistance of steels when alloyed at levels > 8%. It promotes austenite formation at temperatures below the lower critical temperature.
Silicon (Si)	Alloyed in small amounts as a deoxidiser in ferritic steels.
Aluminium (Al)	Used as a grain refiner in steels and a deoxidising agent in triple deoxidised steels.
Niobium (Nb) and Titanium (Ti)	Used to help carbide formation to stabilise stainless steel.

Titanium

Titanium is becoming more widely used in industry because of its excellent properties. These are:

- high strength to weight ratio (strong as steel but half the weight);
- excellent corrosion resistance;
- good mechanical properties at elevated temperatures.

Types of titanium in use are:

- Commercially pure (98–99.5% Ti). May be strengthened by small additions of O_2, N_2, C and Fe and is easily welded.
- Alpha alloys. Mainly single-phase alloys with up to 7% Al and a small amount of O_2, N_2 and C. These can be fusion welded in the annealed condition.
- Alpha-beta alloys. Two-phase alloys formed by the addition of up to 6% Al and varying amounts of beta-forming constituents such as V, Cr and Mo. Can be fusion welded in the annealed condition.
- Ni–Ti alloys that contain a large beta phase, stabilised by elements such as Cr, are not easily welded.

Commercially pure grades and variants of the 6% Al and 4% V alloy are widely used in industry.

Welding problems

There are many problems with welding titanium because it reacts readily with air, moisture, grease, dirt and refractory to form brittle compounds. It is therefore absolutely essential that the weld joint surfaces and filler wire are free of any contamination before and during the entire welding operation. Titanium cannot be welded to most other metals because embrittling metallic compounds that lead to weld cracking are formed.

Above 500 °C, titanium has a very high affinity for hydrogen, nitrogen and oxygen so an inert atmosphere protection must be maintained until the weld metal cools below 426 °C (800 °F). This normally means that a trailing gas shielding is required, i.e. a process utilising inert gases, and welding must be slow enough to enable the trailing gas shielding to be utilised.

- Improper welds might be less corrosion-resistant compared to the base metal.
- Reaction with gases and fluxes makes common welding processes such as gas welding, MMA, FCAW and SAW unsuitable.

Typical imperfections

Titanium welding imperfections are:

- **Porosity**, caused by gas bubbles trapped during solidification. The gas is hydrogen from moisture in the arc environment or from contamination on the filler and/or parent metal surface.
- **Contamination cracking**, caused by iron particles present on the material surface dissolving in the weld metal and reducing its corrosion resistance. They can also cause embrittlement at high iron concentrations. Iron particles present in the HAZ can melt and cause microcracking and corrosion.
- **Embrittlement**, caused by weld metal contamination from

gas absorption (due to poor shielding) or by dissolving contaminants such as dust (iron particles) on the surface.

On the positive side, solidification cracking and hydrogen cracking are not normally found in titanium or its alloys.

A thin layer of surface oxide generates an interference colour at the weld that can indicate whether the shielding was adequate or not. Typical colours are as follows:

- Silver or light straw is acceptable.
- Dark blue may be acceptable for certain service conditions.
- Light blue, grey, white or yellow powders are not acceptable.

Defect avoidance

Titanium is very expensive and many defects can be avoided if the following precautions are taken:

- Avoid steel fabrication operations near titanium components.
- Have dedicated tools used only for titanium.
- Scratch-brush the joint area immediately before welding.
- Do not handle the cleaned component with dirty gloves.
- Cover components to avoid airborne dust and iron particles settling on the surface.
- Maintain adequate shielding gas, trailing gas and purging gas levels.
- In short – keep it clean.

Duplex stainless steel

Duplex stainless steels have a two-phase structure consisting of almost equal parts of austenite and ferrite. They have become very popular because they are approximately twice as strong as common austenitic stainless steels but less expensive, owing to the lower levels of nickel they contain. The characteristic benefits of duplex stainless steels (strength, toughness, corrosion resistance and resistance to stress corrosion cracking) are achieved when there is at least 25% ferrite, with the balance being austenite.

The ferrite in a duplex weld metal is typically in the range of 25 to 60%. In some welding processes utilising a flux, the phase balance of the filler is biased to increased austenite to offset the loss of toughness associated with oxygen pickup from the flux. Thermal expansion and conductivity of a duplex stainless steel are between that of carbon steel and austenitic stainless steel. The operating temperature is normally kept below 300 °C to avoid a degradation mechanism called '475 °C embrittlement'.

Typical defects

- The problems most typical of duplex stainless steels are associated with the heat affected zone (HAZ) rather than the weld metal. The HAZ can suffer from:
  - loss of corrosion resistance;
  - loss of toughness;
  - post-weld cracking.
- The duplex structure is very sensitive to contaminants, particularly moisture.
- Detrimental reactions occur to the material properties if heat input times keep temperatures within the 705 to 980 °C range for too long.
- Rapidly quenched autogenous welds (welds without filler) such as arc strikes and repairs to arc strikes tend to have ferrite levels greater than 60%. These welds can have low toughness and reduced corrosion resistance.

Defect avoidance

- Allow rapid (but not extreme) cooling of the HAZ.
- Limit the temperature of the workpiece because it provides the most effective cooling of the HAZ.
- Limit the maximum interpass temperature to 150 °C (300 °F).
- When a large amount of welding is to be performed, plan

the welding to provide enough time for cooling between passes.

- Avoid PWHT stress relief because duplex steels are sensitive to even relatively short exposures to temperatures in the 300 to 1000 °C range.
- PWHT stress relief in the 300 to 700 °C range may cause precipitation of the alpha prime phase ('475 °C embrittlement'), causing a loss of toughness and corrosion resistance.
- Stress relief in the range of 700 to 1000 °C leads to rapid precipitation of intermetallic phases, resulting in loss of toughness and corrosion resistance.
- Heat treatment of duplex steel, for whatever reason, should be a full solution anneal followed by water quenching.

In summary, the best way to avoid problems with duplex steels is to avoid excessive ferrite levels and limit the total time at temperature in the HAZ.

Material properties

Materials are chosen for service use based on the properties that they possess.

Ductility is the ability of a material to be drawn or plastically deformed without fracture. It is therefore an indication of how 'soft' or malleable the material is. The ductility of steels varies depending on the types and levels of alloying elements present. An increase in carbon, for example, will increase the strength but decrease the ductility.

Hardness is the ability of a material to resist abrasion or penetration on its surface. The harder the material, the smaller the indentation left by an object such as a ball or diamond being impressed upon it. As a general rule there will be a higher risk of cracking as hardness increases.

Toughness is the ability of a material to resist impact (i.e. absorb the energy of an impact). The general rule is that a higher toughness will lessen the risk of cracking.

Heat treatment of steels

Metallic materials consist of a microstructure of small crystals called grains. The grain size and composition help determine the overall mechanical behaviour of the metal. Heat treatment provides an efficient way to manipulate the properties of the metal by controlling the formation of structures, changing the metal properties or controlling the rate of cooling within the microstructure. All heat treatments cycles contain three parts: the rate of climb to the hold temperature (including any hold points), the hold (or soak) time and the cooling rate (see Fig. 4.3).

Heat treatment methods include the use of open flames, electric resistance heating blankets, furnaces and autoclaves. Temperature measurements are taken using indicating crayons (tempil sticks), thermocouples, pyrometers or other methods depending on the level of accuracy required. An inspector should ensure that all heat treatments are carried out in accordance with a specified procedure, make a record of all parameters and ensure that all documents are retained within the quality files.

Annealing

Annealing is a heat treatment carried out to soften and reduce internal stresses on metals that have been work-hardened. The first stage of the annealing process is the

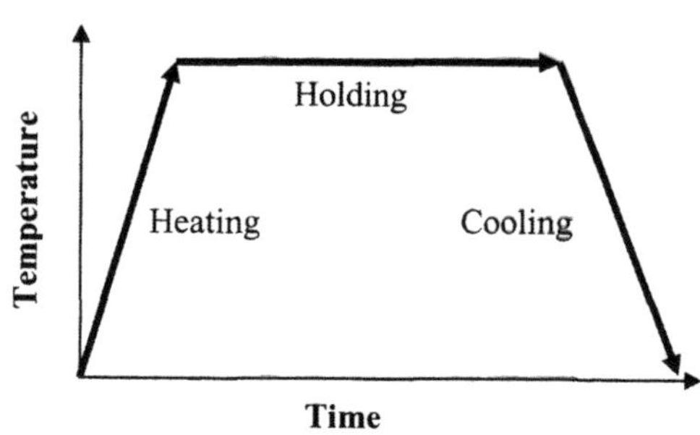

Figure 4.3 Heat treatment cycle

softening of the metal by removal of crystal defects and the internal stresses that they cause. The second stage is recrystallisation, where new grains replace those deformed by internal stresses. Further annealing after recrystallisation will lead to grain growth, which starts to coarsen the microstructure and may cause poor mechanical properties such as a loss of strength and toughness. In work-hardened non-ferrous metals, annealing is used to recrystallise work-hardened grains and the cooling rate is not always critical.

For steel, there are two basic kinds of annealing: full annealing and subcritical annealing.

Full annealing

This is commonly carried out on large castings. The steel is heated to around 50 °C above its upper critical temperature (UCT) and held in a furnace for sufficient time to allow the temperature to become uniform throughout the steel. It is then slowly cooled, causing grain growth. The tensile strength will not be particularly improved but toughness and ductility will increase. The upper critical temperature (UCT) of plain carbon steels ranges from 723 to 910 °C depending on carbon content, so the actual annealing temperature will be dependent upon the carbon content of the steel. Above the UCT the steel structure will be an austenite structure.

Subcritical annealing

Subcritical annealing methods are used to increase the machinability of high carbon steels or for softening worked-hardened mild steels to allow further cold work to be applied. The steel is heated to a temperature above which recrystallisation will take place but below the lower critical temperature (LCT) of 723 °C. This recrystallises the distorted ferrite grains so that the structure becomes softer again. The recrystallisation temperature and time held at temperature will be dependent on the carbon content of the steel.

Annealing generally puts a metal, or alloy, into its most ductile condition. In steels the resultant large grain structure means a reduction in toughness with low impact strength. The lower critical temperature (LCT) is the temperature below which the austenite forms into ferrite and cementite.

Normalising

When an annealed part is removed from the furnace and allowed to cool in air, it is called 'normalising'. It is often used for hardenable steels to regain toughness after high heat input processes have formed large grain structures. The steel is heated to just above its upper critical temperature, held for a specified period and then allowed to cool in air. Small grains are formed, which give a harder and much tougher metal with normal tensile strength rather than the maximum ductility achieved by annealing.

Hardening (quenching)

Hardening of steel is achieved by heating the alloy to above its upper critical temperature until it is a fully austenitic structure and then cooling it rapidly with forced air, oil, water or brine. Upon being rapidly cooled, a portion of austenite (dependent on alloy composition) will transform into martensite. Martensite is very hard and strong but too brittle for most applications. It must therefore be subjected to a process called tempering, which will temper the martensite into a very strong and tough structure. Most applications require that quenched parts be tempered, to impart some toughness and further ductility, although some yield strength is lost.

Tempering

Tempering of steel is used to transform a hard and brittle martensitic structure into a tougher, more ductile structure. There is always a trade-off between ductility and brittleness, and the precise control of time and temperature during the tempering process is necessary to achieve a structure with the correct balance of these mechanical properties. The steel is normally tempered after thermal hardening by heating between 150 and 650 °C (300 and 1200 °F) dependent upon the material properties and the specific mechanical properties required. Tempering can continue up to the lower critical temperature of 723 °C, at which point most of the extra hardness produced by thermal hardening will have been removed, but the fine grain structure produced by the hardening process will remain. Quenched and tempered (QT) steels are normally tempered from between 550 and 650 °C giving them good toughness and strength

Quenching and tempering of 'precipitation hardening' alloys

Precipitation hardening metal alloys have their alloying elements trapped in solution during quenching, resulting in a soft material. Ageing a 'solutionised' metal will allow the alloying elements to diffuse through the microstructure and form intermetallic particles, which fall out of solution and increase the strength of the alloy. Alloys may age naturally at room temperature, or artificially at elevated temperatures. Some naturally ageing alloys can be prevented from age hardening until needed by storing at subzero temperatures.

Precipitation hardening alloys include 2000, 6000 and 7000 series aluminium alloys, some superalloys and some stainless steels. An age hardening alloy can be tempered after quenching by heating at temperatures below the solutionising temperature. During tempering, the alloying elements will diffuse through the alloy and react to form intermetallic compounds. These precipitate out and form small particles that strengthen the metal by impeding the movement of dislocations through the crystal structure of the alloy.

The mechanical properties of an alloy can be determined by careful control of the tempering time and temperature, affecting the size and amount of precipitates. Artificially aged alloys are tempered at elevated temperature, while naturally ageing alloys may be tempered at room temperature. Some superalloys may be subjected to several tempering operations where a different precipitate is formed during each operation. This results in a large number of different precipitates that are difficult to drive back into solution. This contributes to the high temperature strength of precipitation hardened superalloys.

Stress relief

The primary function of post-weld heat treatment (PWHT) stress relief is to relieve internal stresses in welded fabrications by permitting the steel to creep slightly at an elevated temperature. This elevated temperature lowers the material yield point and allows any high residual welding stresses to exceed the new yield stress, causing localised plastic deformation. When the material is cooled down, the yield point increases again but the residual stresses have now been reduced to a lower level.

Stress relief is also used to reduce the hardness of hardenable steels after welding to help prevent brittle fracture. The procedure is basically a form of subcritical annealing and most steels are stress relieved in the 550–700 °C temperature range depending upon their composition. The holding temperature, holding time and cooling rate are critical and must be sufficient to permit the required changes to take place throughout the whole material thickness. For this reason most codes that specify PWHT stress relief will normally specify the minimum holding time in minutes per millimetre thickness (min/mm) or hours per inch (h/in).

Chapter 5

Welding Processes

Welding processes usually utilise heat and/or pressure to form the welded joint. The heat can come from various sources such as an arc, flame, laser beam, electron beam, friction, etc. We will look at the most commonly used arc welding processes in this chapter.

Manual metal arc (MMA)/shielded metal arc welding (SMAW)

Process description

Figure 5.1 shows the equipment used in this process. Fusion is obtained from the heat of an arc formed between a consumable flux-coated electrode and the workpiece. The arc is protected from the atmosphere by a gaseous shroud produced from the melting flux while the weld metal is cleaned of contaminants by the flux, which forms a slag that floats to the top of the weld (Fig. 5.2). This slag must be removed after each weld run before the next pass is added to prevent slag inclusions within the completed weld.

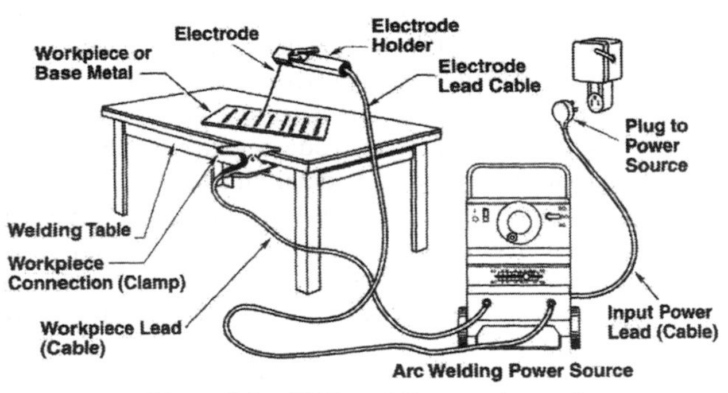

Figure 5.1 MMA welding equipment

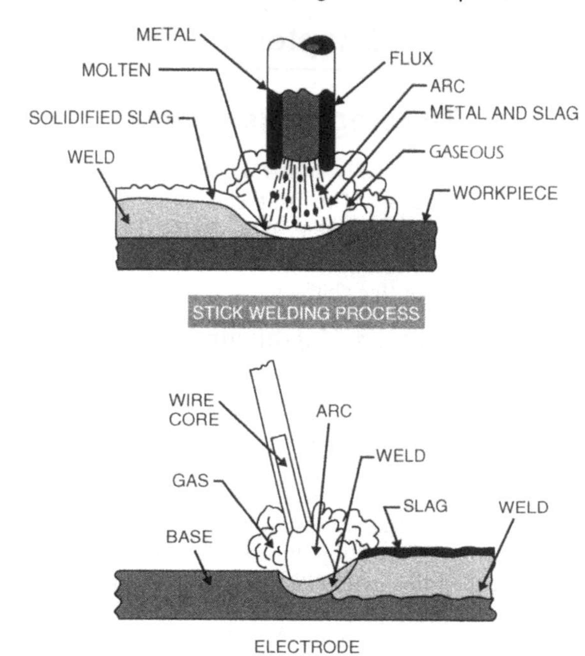

Figure 5.2 MMA welding process

Polarities

Welding can be done by using an alternating current (a.c.) or direct current (d.c.). When direct current is used the welding electrode will be connected to either the positive or negative pole. This is referred to as direct current electrode positive (DCEP) or direct current electrode negative (DCEN). The type of current and polarity is determined by the electrode characteristics. The power source has what is termed a 'drooping' or constant current characteristic (Fig. 5.3). This means that a change of arc length (which controls voltage) will have only a small effect on the welding current as follows:

- As arc length increases, voltage increases and current

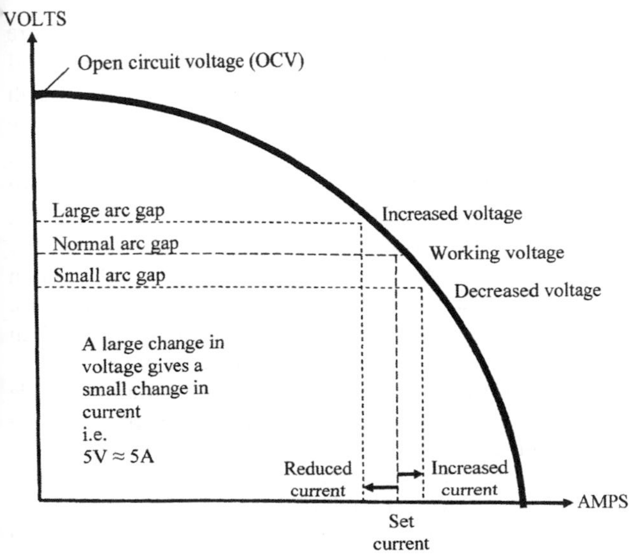

Figure 5.3 Constant current characteristic

decreases. The current decrease is very small and will have little effect on the burn-off rate of the electrode.

- As arc length decreases, voltage decreases and current increases. The current increase is very small and will have little effect on the burn-off rate of the electrode.

Consumables

The only consumables are flux-coated electrodes. These are three main types:

- Basic electrodes. Basic electrodes contain calcium compounds such as calcium carbonate and calcium fluoride in the flux coating. These compounds are helpful for all positional welding as they help produce a fast freezing slag covering on the weld metal. The shielding gas produced from the melting flux to protect the welding arc is mainly CO_2.

Basic electrodes contain low levels of H_2, and are therefore used where high quality welds with good mechanical properties are required (especially high strength welds in high restraint situations, which are susceptible to hydrogen-induced cold cracking (HICC). Basic electrodes require baking at temperatures above 150 °C, storing in an oven at temperatures up to 120 °C, and being used from heated quivers at about 70 °C to ensure they maintain the low H_2 levels required. A modern option is to use electrodes straight from vacuum packs, which give low H_2 levels as long as they are used in accordance with the manufacturer's instructions.

- Rutile electrodes. Rutile electrodes contain titanium oxide in the flux coating. As with basic electrodes the shielding gas produced is mainly CO_2. The rutile coating makes this electrode very welder-friendly due to its ease of use, low fume levels, low spatter levels and smooth weld beads. They are not baked or preheated before use but may be heated for a short period at temperatures up to 120 °C to ensure they are dry before use.
- Cellulosic electrodes. These are electrodes that contain cellulose (which is an organic material) in the flux coating. The shielding gas produced has high levels of H_2, producing a hotter burning arc than CO_2. This hotter arc gives deeper weld penetration and faster welding speeds and is commonly used in the stovepipe welding technique (which entails welding pipes in the vertical down direction).

One problem is that high levels of H_2 are introduced into the weld from the shielding gas, leading to an increased risk of H_2 cracking. To reduce this risk the welding procedures specify that timed hot passes are required; i.e. applying further passes over the root pass while the weld is still hot. This acts as a 'H_2 soak' technique and allows the H_2 to remain as atomic H and dissipate out of the weldment.

Applications

Common uses of MMA are welding of pipelines, nozzles and nodes and medium to heavy fabrications. MMA is commonly used outdoors owing to the good gas shielding supplied from the melting flux.

Typical defects

Typical defects include slag inclusions, porosity, undercut, lack of fusion and profile defects.

Metal inert gas (MIG)/metal active gas (MAG)/gas metal arc welding (GMAW)

Process description

Figure 5.4 shows the equipment. An electric arc is struck between a continuously fed consumable solid electrode wire and the workpiece. The arc is protected by a shielding gas, which can be either inert or active, depending on the material being welded (Fig. 5.5). An inert gas such as argon or helium does not affect the weld pool properties but an active gas such as CO_2 does have an effect. MIG is known as a semi-automatic process because the welding wire is continually fed

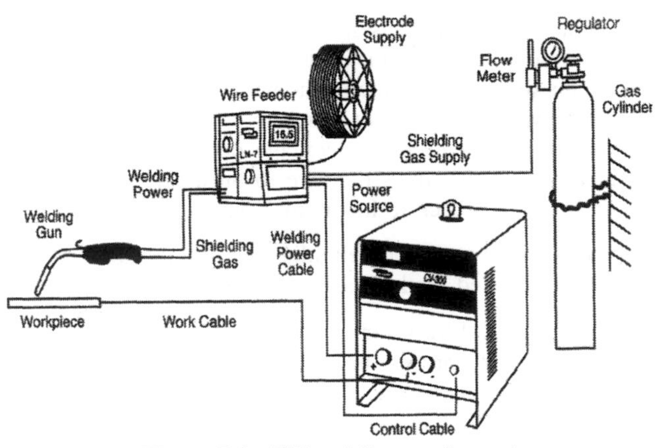

Figure 5.4 MIG welding equipment

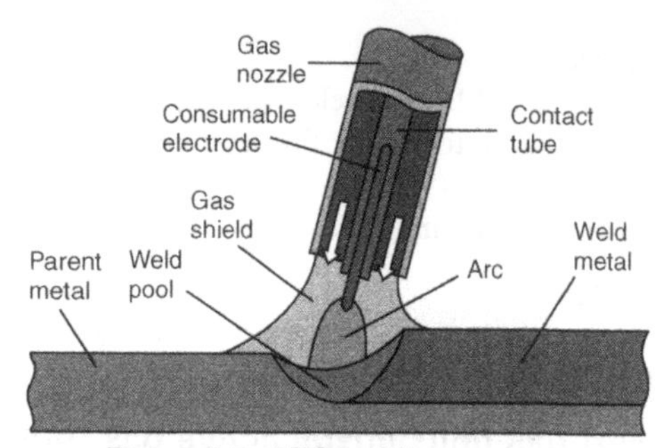

Figure 5.5 MIG welding process

from a reel by the machine but with the travel speed controlled by the welder.

Polarities

MIG/MAG almost always uses a d.c. power source with the polarity electrode positive. The power source has what is termed a 'flat' or constant voltage characteristic (Fig. 5.6). This means that a change of arc length (which controls voltage) will have a large effect on the welding current as follows:

- As arc length increases, voltage increases and current decreases. The current is controlled by wire feed speed and affects the burn-off rate of the wire so the wire will burn off slower and extend out back to its original length.
- As arc length decreases, voltage decreases and current increases. This causes the wire to burn off more quickly until it burns back to its original length.

This is referred to as the 'self-adjusting' arc (because the arc length is adjusted by the machine and not the welder).

Modes of metal transfer

The MIG/MAG process has varying modes of transferring the filler metal across the arc, dependent on what wire feed speed (current), voltage and shielding gas are being used. The

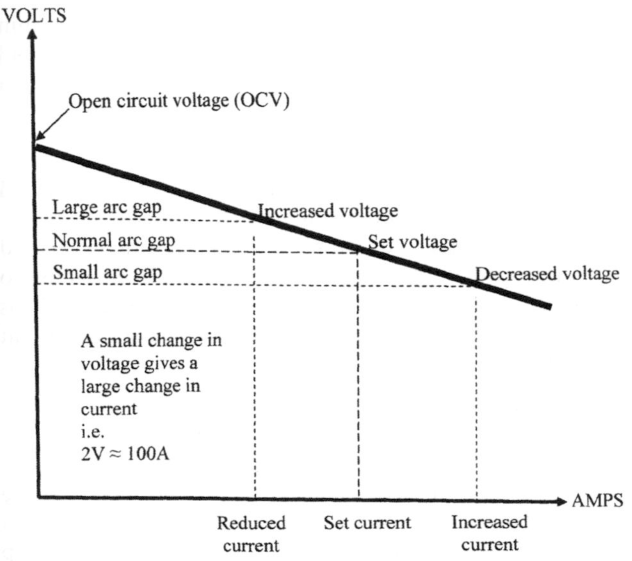

Figure 5.6 Constant voltage characteristic

main modes are short-circuiting transfer, globular transfer and spray transfer.

Short-circuiting (dip) transfer

When voltage and current are low the wire feed speed exceeds the burn-off rate of the wire. The wire 'dips' into the weld pool causing the arc to extinguish and short circuiting takes place. This short circuit increases the current in the wire and the end of the wire becomes molten. A magnetic effect takes place causing the wire to 'neck' and fall off into the weld pool as a molten droplet. The arc then re-establishes and the whole procedure starts again.

In this mode the welding current must be high enough to prevent the wire sticking and the voltage must be high enough to re-establish the arc. Because this mode of transfer has a low heat input it is best suited to the welding of thin materials and for all positional welding due to the small weld pool formed. The downside to this is that lack of fusion can occur in thick section materials.

Globular transfer

Globular transfer takes place between short circuiting and spray transfer modes at medium current and voltage levels. The molten droplets are larger than the wire diameter and some intermediate short circuiting can take place, leading to the arc being unstable and producing high spatter levels. This mode is rarely used except for some filling passes in the flat position.

Spray transfer

Spray transfer takes place with higher currents and voltages. As the current increases there is an increased flow of droplets across the arc and the diameter of the droplets become smaller. The transfer therefore takes place in the form of a fine spray, giving a high deposition rate coupled with deep penetration and a large weld pool. This can lead to difficulty using spray transfer with a thin sheet owing to the risk of burn-through. The large weld pool is also too difficult to control and maintain during all positional welding so it is mainly used with thick sections in the flat or horizontal–vertical positions only. Aluminium can be welded in all positions in the spray mode because the weld pool solidifies quickly, maintaining a smaller more manageable pool.

Pulse transfer

The 'all positional' thickness limitation of the spray transfer mode can be overcome by pulsing the arc to reduce the overall heat input to the work and allow the weld pool to shrink before it gets too large and collapses. This is achieved by regulating the current and voltage to operate in the spray mode for a set period of time, but then immediately reducing them to a level that just keeps the wire tip molten for an equivalent time. An example of this would be to operate on the spray mode for one second, giving deep penetration, but then reduce the amps/volts for one second to allow the weld pool to reduce in size before increasing back up to the spray mode, and so on. In this way the likelihood of getting a lack of fusion-type defects found with the short-circuiting mode is reduced.

Consumables

The only consumables used in a MIG/MAG process are solid wires between 0.6 and 2.4 mm and gases consisting of argon, helium, argon/helium mixtures, CO_2, Ar/CO_2 mixtures, Ar/O_2 mixtures or other proprietary mixtures. It is worth noting the following points in relation to gases:

- Pure CO_2 can be used with steels $\leqslant 0.4\%$ C and low alloy steels using triple deoxidised wire, but it is not usually used in the spray mode.
- Argon produces a better arc in the spray mode and is better with non-ferrous metals and alloys.
- Ar/O_2 (1 or 2%) mixtures are used for stainless steels.
- Helium is normally mixed with argon, oxygen or CO_2. The higher helium contents produce higher arc voltages and heat inputs and give deeper penetrating welds with higher welding speeds.
- Argon/CO_2 (5 to $<20\%$) mixtures are normally used to give a combination of good penetration, a stable arc, less spatter and a flatter weld profile. The lower 5% CO_2 is used in the spray mode and the higher 20% CO_2 is used in the short-circuiting mode. The higher CO_2 level is required to give better penetration in what is a low heat input transfer mode.

Applications

MIG/MAG is commonly used for the welding of structural steels, aluminium alloys and stainless steels. It combines good weld properties with fast deposition rates in light, medium and heavy fabrications.

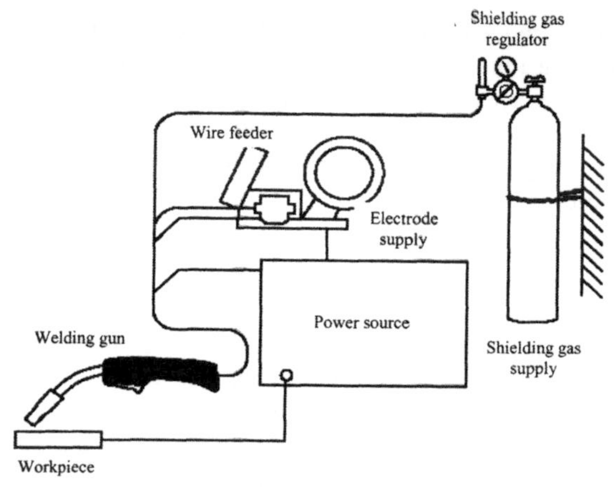

Secondary gas shielded FCAW

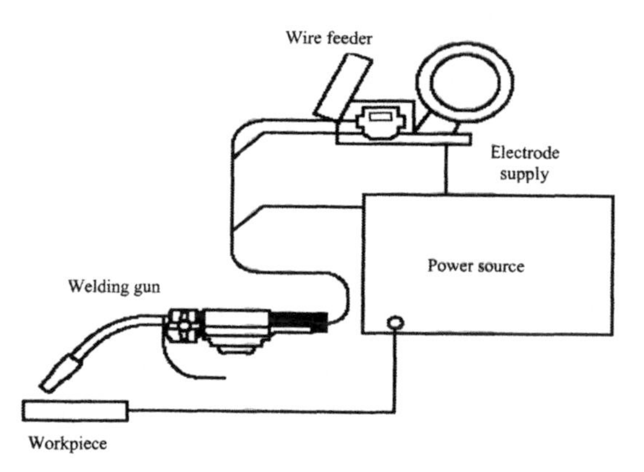

Self shielded FCAW

Figure 5.7 FCAW equipment

Typical defects

Porosity, lack of fusion defects (particularly in the short-circuiting mode), solidification cracking in the spray mode and crater pipes are typical defects.

Flux-cored arc welding (FCAW)

Process description

The equipment is similar to MIG/MAG but uses a flux-cored wire (Fig. 5.7). An arc is formed between a continuously fed tubular electrode wire containing a flux and the work. The arc is protected by a gaseous shroud formed by the flux melting. An external secondary shielding gas can also be supplied through the torch (Fig. 5.8). Using a flux-cored wire enables the addition of alloying elements and the production of a shielding gas more tolerant to outdoor use than MIG/MAG. This means that the benefits of the MMA process can be combined with the speed of the MIG/MAG process. The downside is that it requires slag removal between runs, a backing material for root runs and suitable equipment to remove the large volumes of fume produced from the self-shielding process.

Polarities

DCEP or DCEN is dependent on the wire being used. The power source has a 'flat' or constant voltage characteristic (Fig. 5.6).

Consumables

Cored wires may be self-shielded or gas shielded. Gases may be CO_2, Ar/CO_2 mixtures or Ar/O_2 mixtures. The Ar/O_2 mixture is often used to replace Ar/CO_2 to keep carbon levels at a minimum when welding stainless steels or high alloy materials.

Applications

This process is used in shipyards, structural applications, and other medium and heavy fabrications where positional welding would not be possible with solid wire welding.

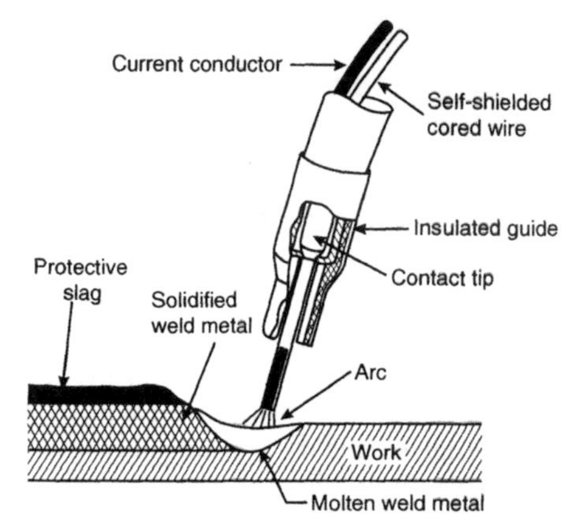

FCAW self shielded

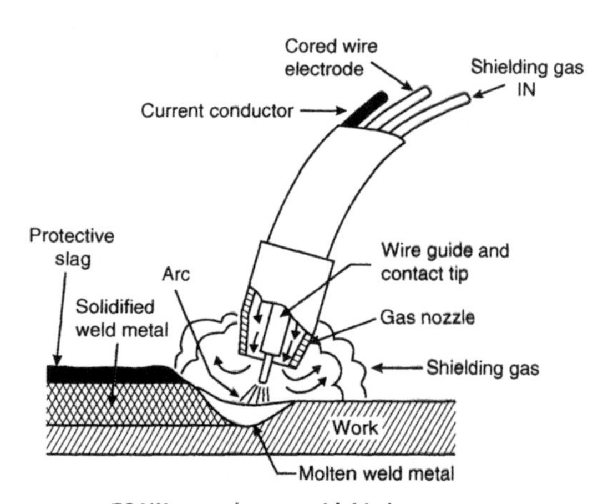

FCAW secondary gas shielded

Figure 5.8 FCAW process

Typical defects

Typical defects are slag inclusions, porosity, undercut, spatter, lack of fusion and profile defects.

Tungsten inert gas (TIG)/gas tungsten arc welding (GTAW)

Process description

Figure 5.9 shows the equipment. Fusion is obtained from the heat of an arc formed between a non-consumable tungsten electrode and the workpiece. The arc and weld pool are protected from the atmosphere by a gas shield supplied through the welding torch. Filler metal can be supplied separately into the weld pool (Fig. 5.10). If a joint is welded without the addition of filler it is called an autogenous weld.

Polarities

Steels are TIG welded using direct current with the welding electrode connected to the negative pole (direct current electrode negative, or DCEN). The reason for this is to keep most of the heat in the workpiece and less in the electrode and so prevent the electrode from overheating and possibly melting.

Aluminium alloys and magnesium alloys are welded using alternating current (a.c.). The positive cycle removes the high temperature oxide layer (known as cathodic cleaning) while the negative cycle helps to keep the electrode from melting.

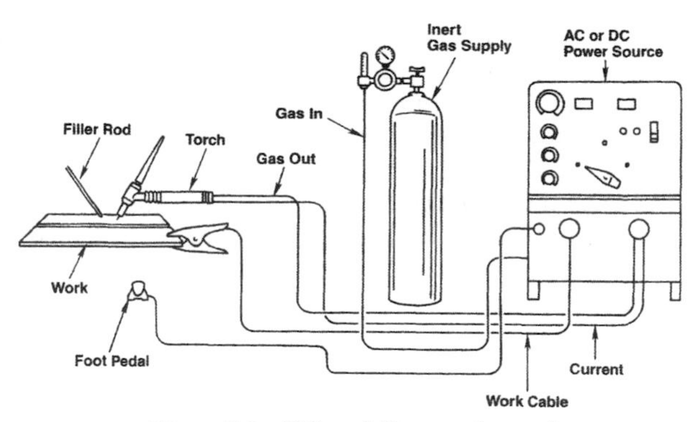

Figure 5.9 TIG welding equipment

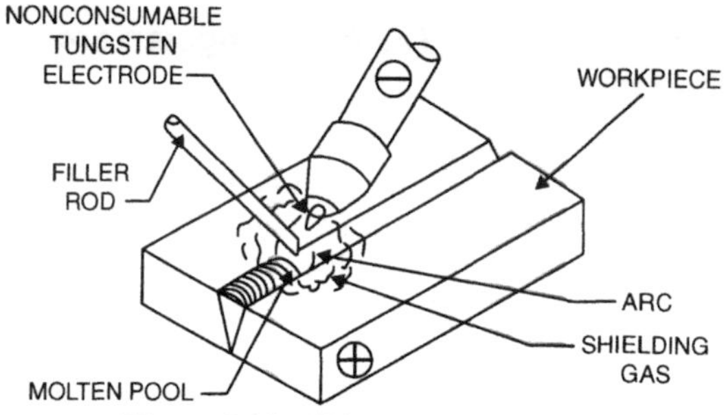

Figure 5.10 TIG welding process

The power source has a 'drooping' or constant current characteristic (see Fig. 5.3).

Consumables

The gas consumables used in the TIG process are mainly inert gases such as argon or helium. Active gas mixtures such as 95%Ar/5%H_2 may also be used in certain applications such as welding stainless steels or nickel alloys. Although this is actually tungsten active gas (TAG), it is generally still referred to with the generic TIG term (except by those trying to sound like they are offering something special).

Filler in wire or rod form is used in the majority of cases to fill the joint, but in some applications a fusible insert (often referred to as an EB after the Electric Boat Company who first supplied them) may be used. This is actually pre-placed filler used for the root run in pipe butt welds, but tends to be used in more specialised applications such as the nuclear industry.

Although the TIG electrode is non-consumable it is often classed as a process consumable owing to the fact that it becomes slowly 'consumed' over time as it is cleaned and shaped by grinding.

Applications

Because manual TIG is such a slow and expensive process it is generally not used on thick materials where high deposition rates are required. On thick materials it is commonly used for the initial root run(s) because the welder has such good control of the weld pool and can therefore achieve a better quality weld than with most other manual processes. Common uses are high quality welds in the aerospace industry, critical root welds in pipe and general light fabrications.

Typical defects

Typical defects are tungsten inclusions caused by touching the electrode into the weld pool during welding, porosity from the loss of gas shielding or surface contamination, oxidation from insufficient purging gas, root concavity from excess purging gas and crater pipes from breaking the arc too quickly.

Submerged arc welding (SAW)

Process description

Figure 5.11 shows the equipment used for SAW. An electric arc is struck between a continuously fed consumable solid electrode wire and the workpiece. The arc is formed and protected within a blanket of flux, which is partially consumed within the process. The flux is supplied from a hopper attached to the weld head and fed through a tube to form a continuous layer in front of the torch deep enough to contain the arc (Fig. 5.12). The weld metal is formed from a combination of the base metal, filler metal and flux constituents, and will therefore be affected by changes to currents or voltages. The weld metal can be cleaned of contaminants by flux additions and then protected by the slag that forms at the top of the weld. This slag must be removed after each weld run before the next pass is added to prevent slag inclusions within the completed weld. Any unused flux can be collected, mixed with new flux and re-used, provided it is not contaminated. Because a flux layer is used, SAW is normally restricted to the flat or horizontal–vertical positions, although some very specialised equipment

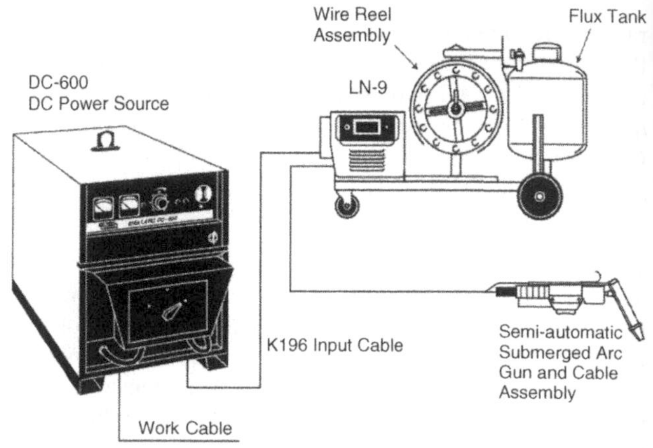

Figure 5.11 SAW equipment

does enable some other positions to be used. Root control is critical and some form of root backing is required.

Polarities

Polarities typically used are as follows.

- Alternating current is used for multihead systems using high currents to prevent magnetic arc-blow problems. Arc-blow is where the magnetic field surrounding the electric arc is affected by another magnetic field causing the arc to deflect. This is not a problem with a.c. because the magnetic field changes direction with each cycle.
- DCEP is used for welding applications to give deep penetration.
- DCEN is used for surfacing or cladding applications (DCEN gives shallow penetration).
- Multihead systems of two or more heads often use combinations of polarities to increase penetration and deposition rates without causing arc-blow. It is common for the leading head to be d.c. while the trailing head(s) are on a.c. The power source normally has a ‘flat’ or constant

voltage characteristic (Fig. 5.6) but can have a 'drooping' or constant current characteristic (see Fig. 5.3) in certain applications.

Consumables

Consumables consist of reel-mounted bare wire or flux-cored electrodes and granular fluxes.

Fluxes

The main types of SAW fluxes are agglomerated or fused.

Agglomerated fluxes are of light colour, globular and of irregular size. They are easily crushed and feel dry and smooth to the touch. Deoxidisers and additional alloying elements are added during manufacture. They are also hygroscopic

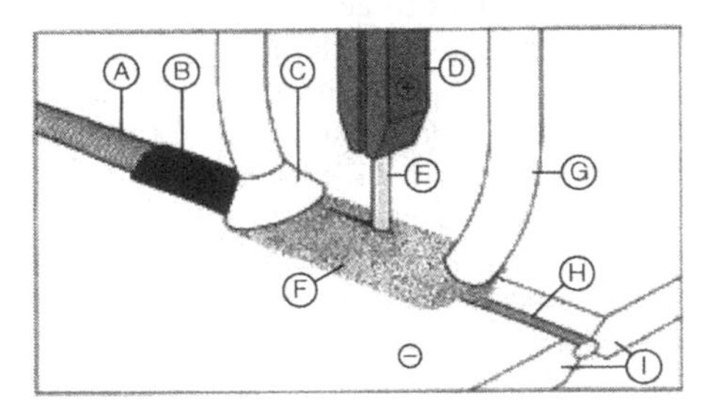

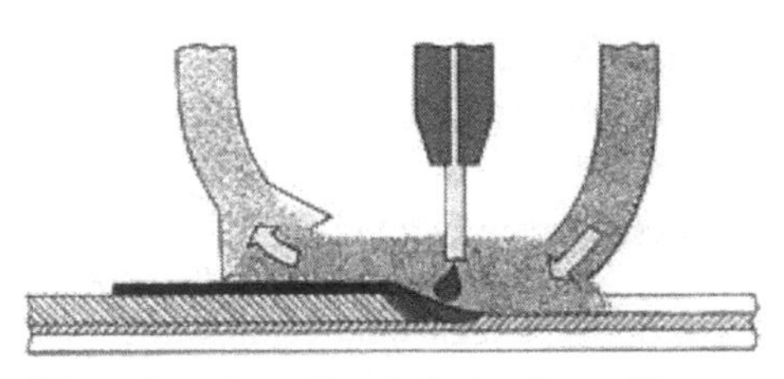

Schematic representation of submerged-arc welding

A Finished weld
B Slag
C Powder removal
D Electrode holder
E Filler wire
F Flux
G Flux supply
H Root bead
I Parent metal

Figure 5.12 SAW process

(absorb moisture) and therefore must be kept dry and pre-heated before use to remove the moisture. Agglomerated fluxes give better mechanical properties to the welded joint than fused fluxes and are used in low hydrogen applications.

Fused fluxes are formed by melting their constituents together at high temperature and then cooling them to form a glass-like flux. The granules are solid and feel like grains of sand when rolled between the fingers. They are moisture resistant, easy to use, give good weld profiles and have good slag detachability, but the weld metal properties are not as good as those using agglomerated fluxes. These are ideally used in general purpose applications.

Fluxes are also classified as basic or acidic, which refers to the ratio of basic oxides to acidic oxides that they contain. The higher the basicity then the greater the flux moisture absorption and related difficulty in removing it. Agglomerated fluxes will be basic whereas the fused fluxes will tend to be acidic. An increase in basicity will increase toughness but reduce arc stability, reduce weld profile and make slag removal more difficult. The flux with the highest basicity giving good arc stability, acceptable weld profile and slag removal should normally be chosen.

Applications

Because of the deep penetration and high deposition rates achievable with SAW it is commonly used in heavy fabrication environments such as shipbuilding and pressure vessel manufacture.

Typical defects

Typical defects associated with this process are shrinkage cavities formed where high weld depth/width (>2:3) ratios are present, solidification cracking caused by high dilution levels with parent material or a high depth/width ratio, porosity due to damp fluxes or insufficient flux depth, hydrogen cracking from damp flux and lack of fusion from arc-blow or an incorrect technique.

Chapter 6

Non-destructive and Destructive Testing

Liquid penetrant testing (PT)

Liquid penetrant examination, often called dye penetrant or penetrant testing (PT), is used to find surface breaking defects only. It involves the use of a cleaner (degreaser), a liquid penetrant and a developer. The most common PT system on site involves the use of these three materials from cans and is referred to as the 'three can system'. Figure 6.1 shows the idea.

A typical colour contrast procedure involves preparing the surface to remove any spatter, slag or other imperfections that could retain the penetrant and mask relevant indications. The surface is then thoroughly cleaned, using the cleaner, to remove any surface oil or grease, which could prevent the red liquid penetrant being drawn into surface breaking cracks or indications. It is then dried using air or lint free cloths.

The penetrant is then applied by spray or brush and left for a dwell time as specified in the procedure. This dwell time must be long enough to enable the penetrant to be drawn into surface breaking defects by capillary action. Excess penetrant is then removed using cloths dampened with the cleaner. The cleaner must not be sprayed directly on to the component otherwise penetrant could be washed out of relevant indications.

Developer (a white chalk like substance) is then lightly sprayed on to the surface causing the red penetrant to be drawn out of any indications by reverse capillary action and the blotting effect of the developer. Any indication highlighted by the red penetrant against the white developer can then be assessed. In effect, PT is a form of enhanced visual

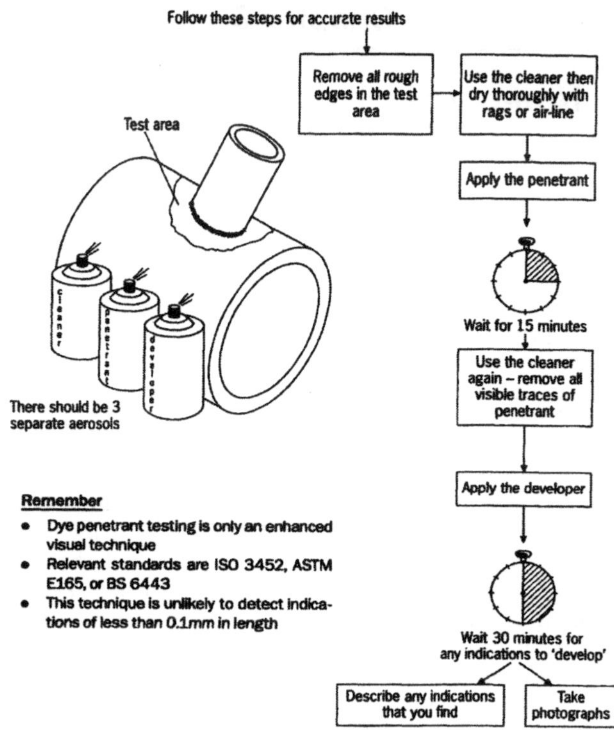

Figure 6.1 Liquid penetrant testing

technique that is often used to confirm visual uncertainties. Viewing is done under good lighting conditions of 1000 lux (100 ft candles).

Typical colour contrast penetrants used are solvent removable, post-emulsifiable or water washable. Fluorescent penetrants can also be used in solvent removable, post-emulsifiable or water washable form but are viewed under ultraviolet (UV) light and tend to be more sensitive than the colour contrast methods. Viewing is done in a darkened room with a light intensity of 1000 W/cm^2 on the surface being viewed.

Advantages

- Can test most materials including non-magnetic ones.
- Cheap and simple to use.

Disadvantages

- Can only find surface breaking defects.
- Good surface preparation and cleaning is required.

Magnetic particle testing (MT)

Magnetic particle testing is used to find mainly surface breaking defects in ferromagnetic materials. Sometimes it is possible to find slightly subsurface defects when used with a permanent magnet or d.c. electromagnet. A magnetic flux (or field) is introduced into the material and any defects cutting across the magnetic flux can be detected when ink or powders containing ferromagnetic particles (iron filings) are applied to the material. What happens is that a flux leakage occurs at the defect, which effectively makes the defect a magnet in its own right. This 'magnet' attracts the ferromagnetic particles, which take the shape of the defect. Figure 6.2 shows the arrangement.

The magnetic flux can be introduced from:

- a permanent magnet;
- an electromagnet (either a.c. or d.c.);
- electric prods (either a.c. or d.c.) between which a current flows (and the current flow is surrounded by a magnetic field).

Ferromagnetic particles can be applied to the material as:

- a black ink (viewed against a pre-applied white contrast paint);
- fluorescent ink (viewed under UV light conditions);
- red or blue dry powders (used at higher temperatures).

The magnetic flux is applied in two mutually perpendicular directions. The reason for this is that if the flux direction runs parallel to the defect then magnetisation of the defect will not take place and the ferromagnetic particles will not be attracted to it. Testing in two directions ensures that the flux cuts any linear defects by at least 45 degrees. The magnetic field strength can be checked using a checker such as a 'pie gauge' or 'burma castrol strip'.

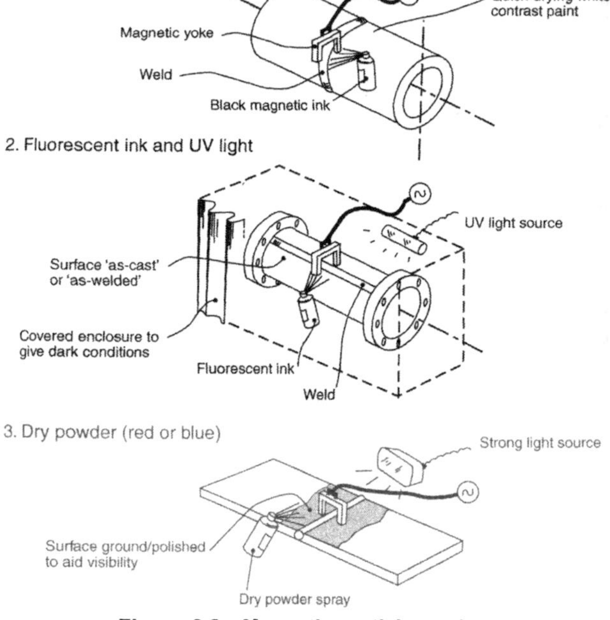

Figure 6.2 Magnetic particle testing

Advantages

- Quick and simple to use.
- Can find slightly subsurface defects (under certain conditions).
- Portable when permanent magnets used.

Disadvantages

- Can only be used on ferromagnetic materials.
- Cannot be used on austenitic stainless steels (non-magnetic).
- Electric prods can cause arc strikes.
- The component may need demagnetising on completion of testing.

Ultrasonic testing (UT)

Ultrasonic testing is used to find internal defects within a weld or body of a component being tested. A probe emits a sound wave that is passed through the material. If this sound wave hits a defect then all or part of it gets rebounded back to a receiver in the probe and the size and position of the defect can be plotted on a graph by a skilled operator. Figure 6.3 shows the arrangement.

Angled probes send the wave into a weld at angles suitable for the weld preparation bevel angles used. A zero degree 'compression' probe is used first to check for any laminations in the parent material that could deflect angled waves and mask defects in the weld.

The surface of the component must be clean and smooth and a couplant applied to exclude air from between the probe and component. The couplant must be suitable for use on the material being tested and then be thoroughly cleaned off afterwards to prevent any risk of corrosion or degradation of the component in service. Wallpaper paste is often used as a couplant because when it dries it can be easily peeled off.

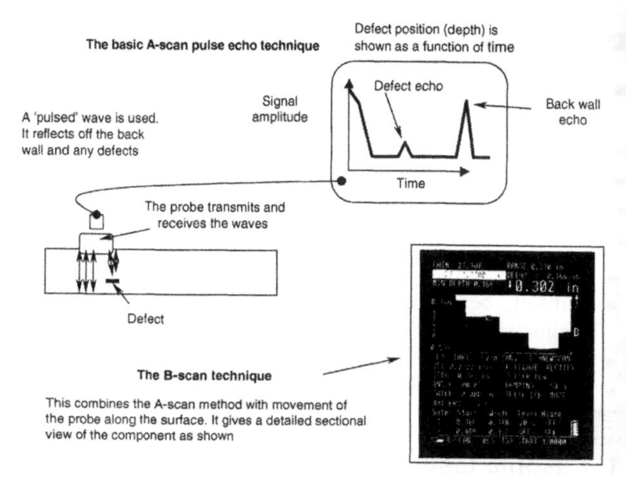

The C-scan technique

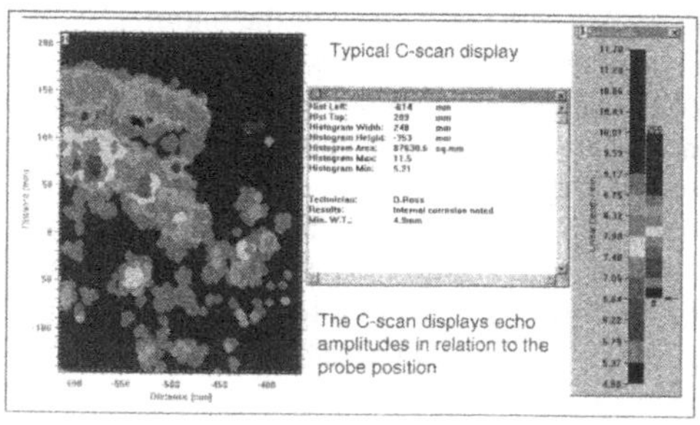

Figure 6.3 Ultrasonic testing

Advantages

- Can find linear type defects in most orientations.
- Is portable.
- Is safe and therefore does not require production to be stopped or the area cleared.

Disadvantages

- Surfaces must be smooth and free of all spatter and arc strikes.
- Very dependent on skill of the operator.
- Normally limited to materials above 6 mm thick.
- Basic technique does not give a permanent record.
- Not so good on coarse grained materials.

Radiographic testing (RT)

Radiographic testing is used to detect internal defects in welds. Although it can find planar (two-dimensional) defects such as cracks or lack of fusion it will not find them in all orientations. It will, however, more easily find volumetric defects such as porosity or slag inclusions or shape defects such as undercut or excess root penetration. It can also be used for profile surveys of pipework and components to check for loss of wall thickness caused by corrosion and/or erosion. Figure 6.4 shows the basic technique.

Gamma rays (from a radioactive isotope) or X-rays (from a machine) are passed through the material and strike a film causing it to darken. The film gets darker the more radiation that hits it so volumetric defects such as porosity that allow more radiation through the material will show as areas darker than the surrounding area. Conversely, areas such as excess penetration, where more radiation is absorbed by the material, will show as lighter than the surrounding area. Gamma radiography is commonly used on site because it is portable and does not require a power source. It is inherently dangerous, though, because the radioactive isotope (source) cannot be 'turned off' and is always radiating. Storage, transportation and use of the source must therefore be closely controlled to ensure the safety of the radiographers and workplace personnel.

The type of isotope used will depend on the thickness of material to be tested. The most common gamma isotope is iridium 192 but thinner materials may use other isotopes

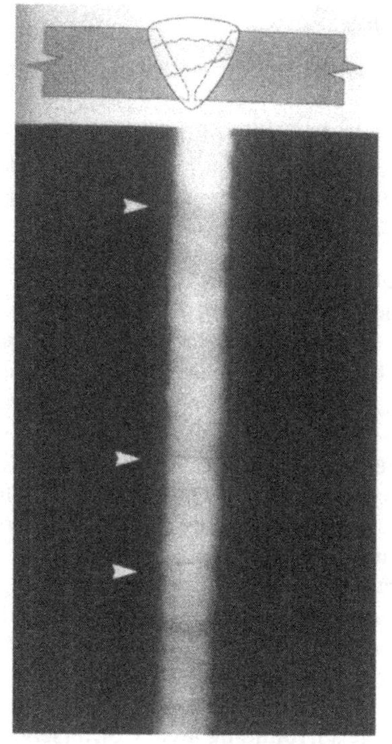

Figure 6.4 Radiographic testing

such as yturbium and (for thicker materials 40 mm+) cobalt 60. Cobalt 60 is incredibly dangerous and should only really be used when iridium 192 is impracticable. It is a good idea to finish having all the children you want before using cobalt 60 or, better still, get someone else to use it and keep well back. On a more serious note – treat it with great respect. The sources are contained in carriers made from depleted uranium and are only wound out for the required exposure time.

Quality of the radiographic film is measured using four main parameters:

1. **Density** is a measure of how much light passes through the film. The higher the density number, the darker the film. Normal acceptance levels are densities of between 1.8 minimum to 4.0 maximum for X-ray and 2.0 to 4.0 for gamma, but it varies between standards. The density measurement is normally taken along the area of interest (i.e. the weld length) using a piece of equipment called a densitometer.
2. **Sensitivity** is an indication of the smallest defect that can be seen on the image. A wire type or hole type image quality indicator (IQI) is used to determine the smallest defect visible on the image. The wire type IQI is the more common type used and consists of either six or seven wires depending on the standard used (see Fig. 6.5). The American ASTM standard specifies six wires but the European standard specifies seven. Sensitivity is expressed as a percentage derived from the thinnest wire visible divided by the material thickness. The acceptable percentage will vary depending on material thickness. To save calculations ASME V and EN standards give tables specifying the smallest wire that

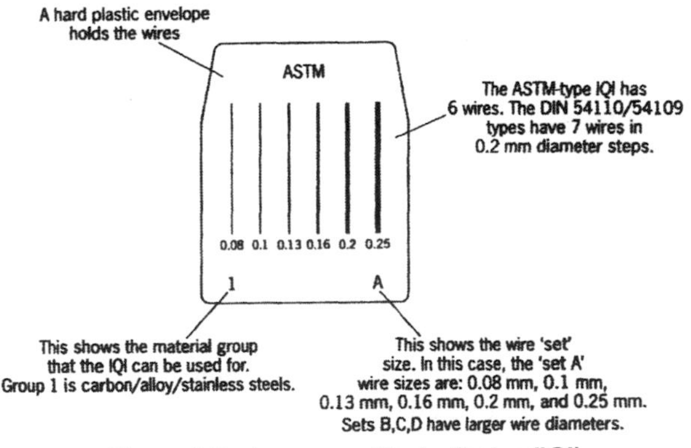

Figure 6.5 Image quality indicator (IQI)

must be visible to give the correct sensitivity value for different material thicknesses and different radiographic techniques.

3. **Geometric unsharpness (Ug)**, also known as penumbra, is a measure of the 'fuzziness' of the radiographic image. Geometric unsharpness (Ug) is calculated from the formula

   $\mathrm{Ug} = Fd/D$
   where
   F = effective source size or focal spot
   d = object to film distance
   D = source to object distance

   Acceptance is based on the relevant code requirement. ASME V Section 2 gives recommended maximum limits for Ug ranging from 0.020 in (0.51 mm) for material thicknesses below 2 in to 0.070 in (1.78 mm) for materials greater than 4 in.

4. **Backscatter** is stray radiation that can expose the film. A lead letter 'B' is placed on the back of the film and if excessive backscatter is present a light image 'B' will be visible on the film, which should then be rejected. A darker image, on the other hand, is not a cause for rejection.

Advantages

- Gives a permanent record.
- Surface condition is not so critical.
- Lagging does not require removal.

Disadvantages

- Safety is an issue.
- May require production to be shut down.
- Personnel must be excluded from the vicinity.
- Will not find linear defects in all orientations.

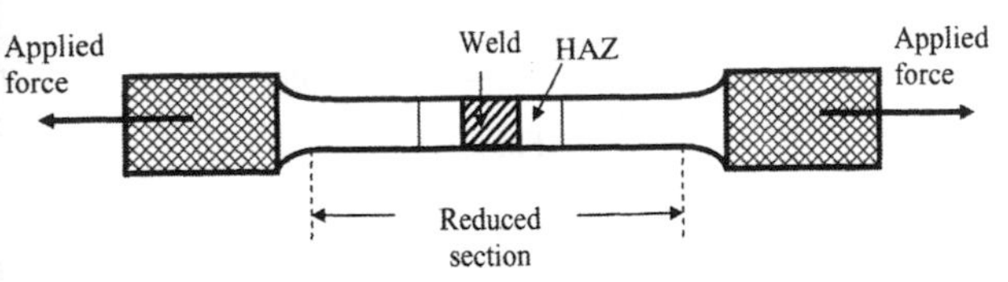

Figure 6.6 Tensile specimen: reduced section transverse test

Tension tests

The main objective of a tension test on a welded joint is to measure the yield and tensile strengths of the sample under test. A pulling force (load) is applied to the sample and the yield and ultimate tensile strength (UTS) are measured and recorded on the test form along with the material type, specimen type, specimen size and location of the fracture. Weldments are normally subjected to a reduced transverse tensile test (see Fig. 6.6).

Manufacturers of welding consumables carry out a longitudinal 'all weld metal' tensile test to measure the tensile strength, yield point and percentage elongation ($E\%$) of the deposited weld metal. The sample is taken from the centre of the weld and consists of weld metal only.

Bend tests

Bend tests give an indication of weld quality and a rough indication of ductility by putting the weld and HAZ under tension. The weld and HAZ must be included within the bent portion and a limit given to the size of linear openings permitted on the surface under test. Bend tests are normally transverse tests taken across the weld and include the weld, HAZ and base material. Longitudinal tests are more unusual but can be used where dissimilar base materials with widely differing properties are welded or where the weld metal is of greatly differing properties to the base metal. Guided bend tests are those that have the sample bent into a guide (see Fig. 6.7).

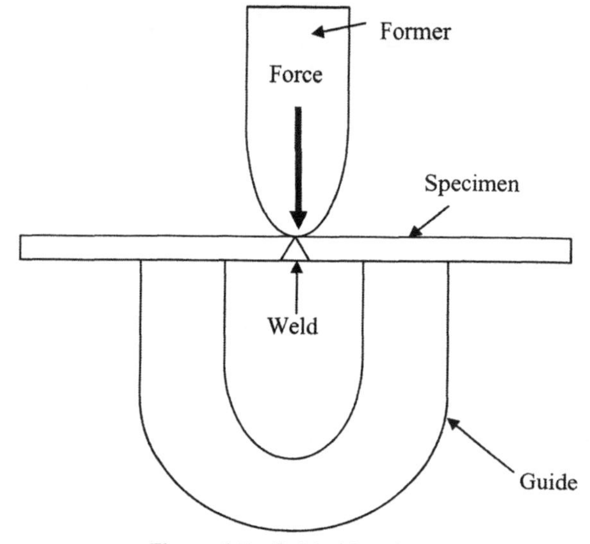

Figure 6.7 Guided bend test

Bend tests are classed as face, root or side bends (see Fig. 6.8) depending on which surface is under tension; a face bend will have the face under tension while a root bend will have the root under tension. A side bend test is used where the sample is too thick to form a face or root bend. With a side bend test the cross-section of the whole weld is put under tension and checked for internal defects such as lack of sidewall fusion or inter-run fusion. It is worth keeping in mind that the side bend is just a snapshot of the weld at one particular point within the weld length.

Charpy tests

Material impact toughness can be measured by various types of test such as the Charpy V-notch impact test, Izod test or K_{IC} test. The most commonly used test is the Charpy impact test (see Fig. 6.9), which gives an indication of the toughness of a material at a specified temperature. It is not a

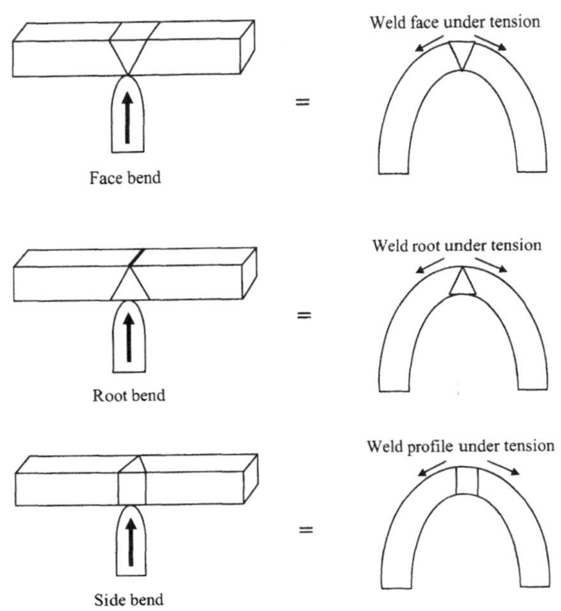

Figure 6.8 Types of bend test

particularly accurate test but can give a general indication of the ability of a material to resist brittle fracture at its minimum design material temperature. The test consists of holding a machined specimen, of a specified size (normally 55 mm × 10 mm × 10 mm) containing an accurately machined notch of specific dimension, at both ends as a simple beam. A pendulum impacts on the specimen and the start and finish heights of the pendulum are measured. The difference in height equates to the energy absorbed by the specimen before it fractures. This absorbed energy is usually measured in joules on a scale attached to the machine.

Three Charpy specimens are tested at each specified temperature and the final result taken as an average of the three. Tests can be done at various temperatures and recorded in a graph to form the Charpy curve (see

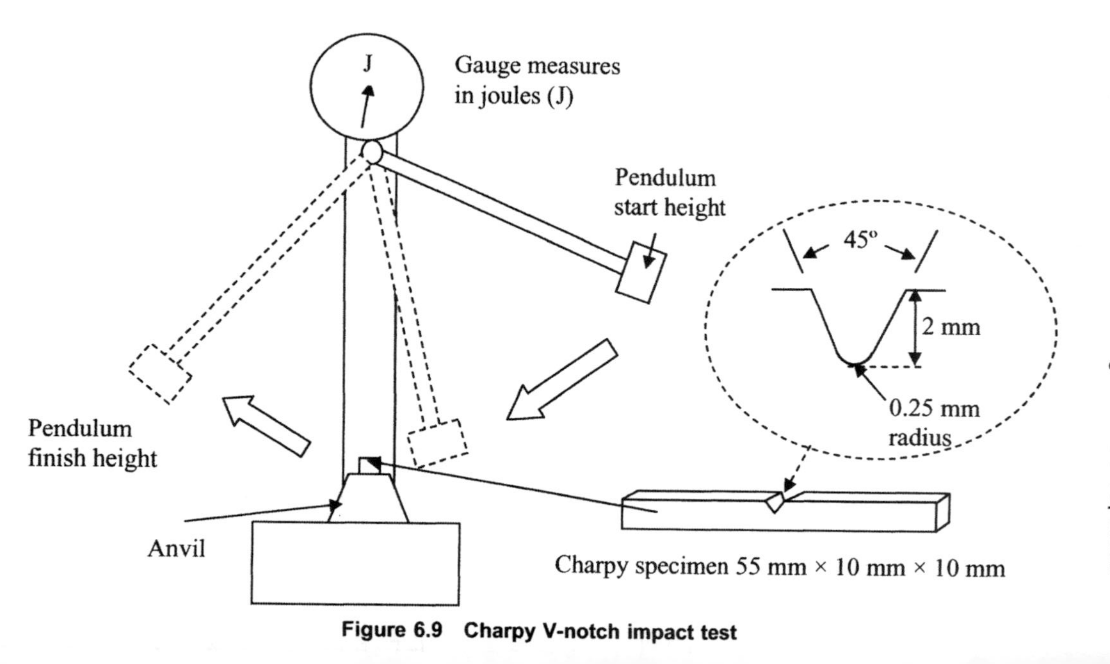

Figure 6.9 Charpy V-notch impact test

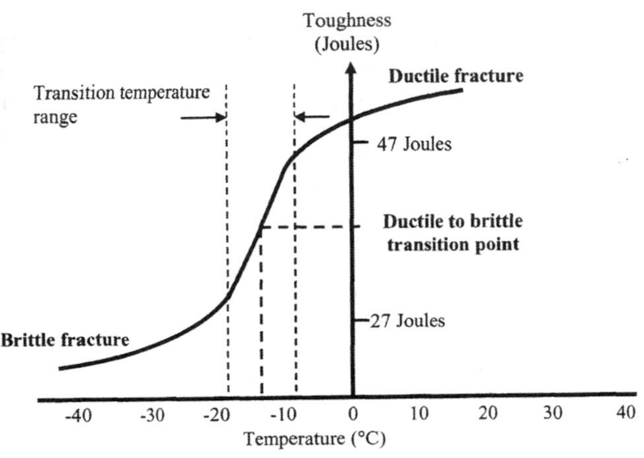

Figure 6.10 A typical Charpy curve

Fig. 6.10) and determine the ductile-to-brittle transition temperature. The ductile-to-brittle transition temperature is the temperature at which the test specimen will start to become more brittle than ductile. You would therefore not want to use the material at design temperatures below this as it would have an increased risk of failing in a brittle manner. Remember that this is not an accurate test reflecting the material behaviour under actual service conditions, so the results should therefore be used with caution.

A more accurate test to check a material's likelihood of failing in a brittle manner is the crack tip open displacement (CTOD) test, sometimes referred to as a K_{IC} test.

Hardness testing

The most common hardness tests are Vickers, Rockwell and Brinell. Hardness is defined as the ability of a material to resist indentation on its surface. Hardness tests consist of impressing a ball (Brinell or Rockwell) or diamond shape (Vickers or Rockwell) into the material under a specified loading and measuring the width of the indentation to give a relative hardness reading (Fig. 6.11). The smaller the width of the indentation the harder the material. Hardness testing normally encompasses the weld and HAZ and is usually done to confirm that PWHT has been carried out correctly. When hardness testing is done for weld procedure qualification purposes it is often done through the weldment thickness as hardness levels can vary considerably through the thickness.

The various types of hardness testing have their own units:

- Vickers test: HV (Vickers hardness);
- Brinell test: HB (Brinell hardness);
- Rockwell: HR (Rockwell hardness).

A Shore Schlerescope is a portable dynamic hardness test using equipment similar in size to a ballpoint pen. It drops a weight from a height on to the test surface and measures the height of the rebound. The higher the rebound the higher the hardness value, which can be read off in any selected unit. It may be used by the welding inspector to gauge hardness values on site, but the accuracy depends on the condition of

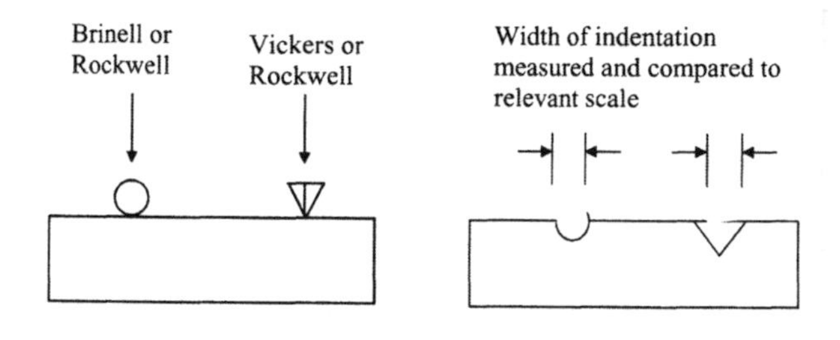

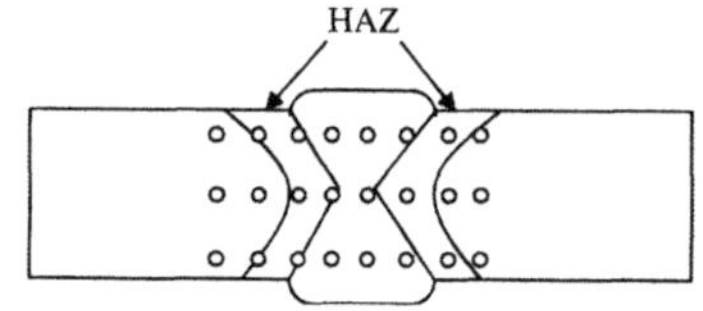

Hardness measurements taken through thickness on a macro sample

Figure 6.11 Hardness tests

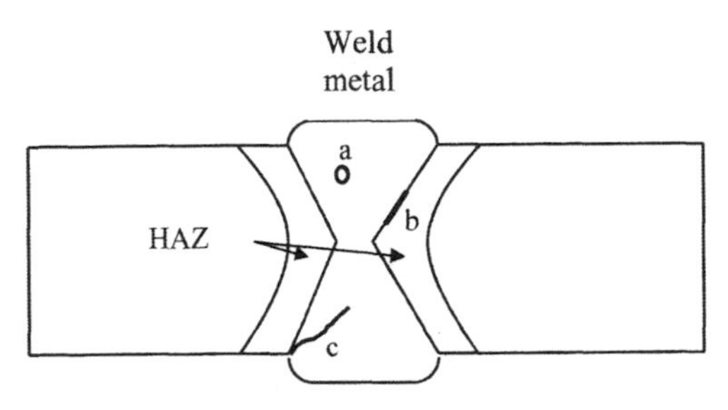

Double V butt macro sample showing

a. A pore
b. A lack of sidewall fusion
c. A crack

Figure 6.12 Macro sample

the test surface and the support of the test piece during the test.

Macro samples

Macro tests (Fig. 6.12) are used to check the internal weld quality of a welded test coupon and so are used in welder performance qualification tests (welder approval tests). A section is removed from the test coupon and then the surface to be inspected is made smooth and polished by lapping to a finish of approximately 600 grit before etching it with 'nital' (a mixture of around 10% nitric acid in industrial alcohol). The mixture will depend on the material being tested.

The macro specimen is then inspected at a magnification of ×5–×10 for welder-induced imperfections such as porosity, slag inclusions, cracks, lack of fusion etc. It is common for the macro to be taken at a stop/start location (particularly in the root runs) as this is where defects are most likely to be found. Obviously any stop/starts in root or hot passes would have to be marked on the sample by an inspector before subsequent runs obscured them.

The sample should also be inspected when it is first removed to check for any imperfections which may then be

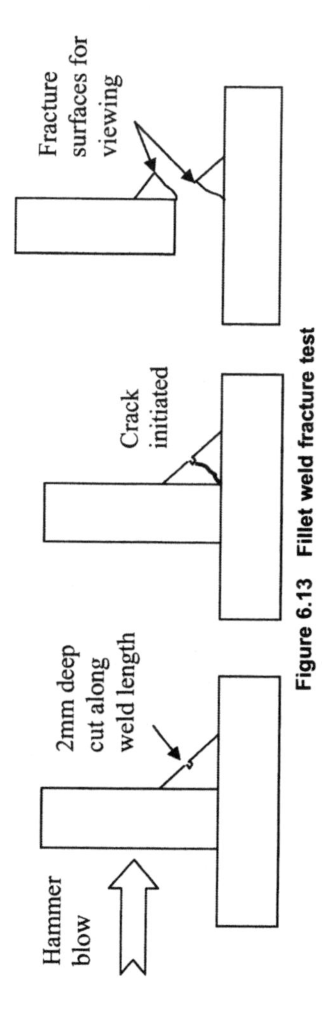

Figure 6.13 Fillet weld fracture test

removed during subsequent preparation. It is common for macro samples to be varnished and mounted in plastic to preserve them and avoid direct handling of the prepared surface.

Micro samples

A micro sample is prepared to permit examination of the grain structure under much higher magnifications than that used for macros. Magnification is normally in excess of ×100 and can be up to ×2000+ using electron microscopes. The sample is prepared in a similar manner to a macro but the surface is finished with diamond paste to give a much more polished finish of around 1–3 μm before etching with a nital solution (see macro preparation). Micro samples are normally used for research purposes or investigations into the cause of defects or failures.

Fillet fracture test

Fillet fracture tests are used for welder approval testing using fillet welded test coupons. A typical test coupon will range from 150 mm (6 in) to 300 mm (12 in) long and will ideally contain a stop/start in the centre. The end 25 mm (1 in) will be removed either side and discarded (although it is a good idea to make one or both cut ends into macros). The central vertical section will be loaded in the direction shown in Fig. 6.13 until the specimen either fractures or bends flat. A common method of applying loading is to strike it sharply with a hammer in the loading direction. Fracture can be assisted by cutting a groove 2 mm deep along the centre of the weld to act as a fracture initiation point. The fracture surfaces can now be inspected for any internal defects such as porosity, slag inclusions or lack of fusion defects. The root can also be inspected to ensure root penetration along the full length has been achieved.

Butt weld fracture test (nick break)

Another welder approval test is the nick break test (Fig. 6.14), which is similar to the fillet fracture test but is used to

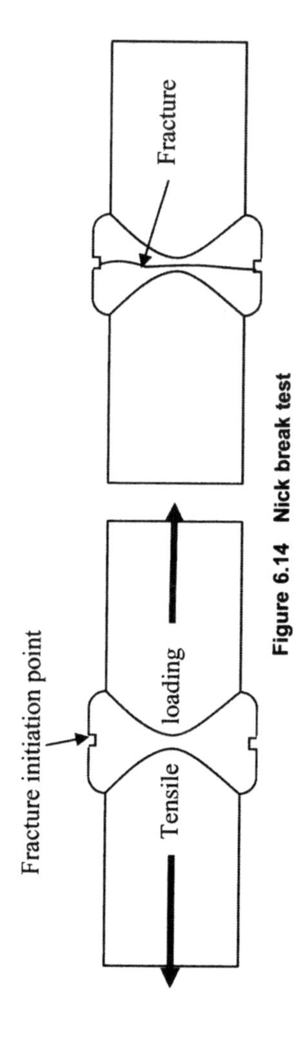

Figure 6.14 Nick break test

assess a butt weld. The weld has a notch cut along the weld length and is then either put under a tensile loading until fracture occurs or is placed in a vice and fractured with a hammer blow. The fracture faces are then inspected for internal imperfections.

Chapter 7

Fracture Modes and Welding Defects

Imperfection categories

Imperfections can be broadly classified as those formed during fabrication and those formed during service. Fabrication imperfections include:

- Cracks
- Porosity and cavities
- Lack of fusion
- Solid inclusions
- Profile imperfections
- Wrong sizes

Service failure modes include:

- Brittle fractures (cracks)
- Fatigue fractures (cracks)
- Stress corrosion cracking (SCC)
- Re-heat cracking
- Creep failure

Cracks

Cracks can be classified by shape (longitudinal, transverse or branched) and position (HAZ, base metal, centreline, crater). An inspector will rarely classify a crack as a particular type (i.e. fatigue, HICC or SCC) because in most cases it will not be possible to determine the precise cause of a crack until an examination is carried out. The shape and position are facts, anything else is supposition.

Hydrogen-induced cold cracking (HICC)

HICC may occur in the HAZ of all hardenable steels (i.e. C, C/Mn) or in the weld metal of high strength low alloy (HSLA) steels that are microalloyed with small amounts of titanium, vanadium or niobium (typically <0.05%). The

hydrogen breaks down at increased temperatures into atomic hydrogen (which has a small atomic size) and escapes to the atmosphere through the steel microstructure. When the temperature reduces to below around 300 °C the hydrogen starts reforming to the hydrogen element and will no longer be able to escape from the material. As the H_2 reforms it may build up an internal pressure stress within the material structure itself.

A ductile metal structure can absorb this H_2 without penalty as the metal is able to plastically deform. If, however, the metal is of a hardened (i.e. martensitic) structure less able to deform, then the stress level built up in the material may be enough to cause a fracture to occur. The four critical factors involved in a hydrogen crack are:

- A hydrogen content of > 15 ml/100 g weld metal. The hydrogen comes from moisture, paint, oil, grease, damp electrodes or fluxes, loss of shielding gases or cellulosic electrodes (H_2 is the shield gas).
- A stress level > 50% yield. The stress comes from residual welding stress, restraint stress, etc.
- A hardness > 350 Vickers. The hardness refers to a crack-sensitive microstructure and is related to the Cev of the steel and the formation of martensite, a hard structure caused by rapid cooling of steels.
- A temperature < 300 °C. There is nothing that can be done to stop the temperature eventually falling below the critical level on completion of the welding and any required PWHT, so NDT needs to be carried out for up to 72 hours after welding to check for delayed cracking.

All of the four factors must be present at the same time for a crack to occur, so if any factor is reduced below its critical level a crack will be avoided. Each factor can be reduced as follows:

Reduce the hydrogen level by:

- use of correctly heat treated low H_2 electrodes;
- removal of moisture, oil, grease, paint, etc., from materials;
- use of a low H_2 welding process;
- carrying out a hydrogen soak.

Reduce the stress level by:

- removal of restraints;
- ensuring a good joint fit-up;
- carrying out PWHT stress relief.

Reduce the hardness by:

- preheating the joint;
- carrying out PWHT.

Carry out delayed NDT:

- The temperature will eventually fall below the critical level but a crack may not occur until days afterwards, so carry out delayed NDT.

Solidification cracking

Solidification cracking (also known as centreline cracking or hot cracking) is a fracture that occurs in the weld metal of ferritic steels with a high sulphur or phosphorus content or in joints with a large depth/width ratio.

It occurs in ferritic steels with a high sulphur content because during welding the sulphur joins with iron to form iron sulphide (FeS). This FeS has a lower melting point than steel and therefore remains as a liquid as the steel solidifies. The centre of the weld is the last place to cool so this liquid FeS is pushed to the centre of the weld and forms a liquid film on the grain boundaries, causing a lack of adhesion between the grains. The weld metal contraction due to cooling leaves a high tensile stress, which can pull the weld

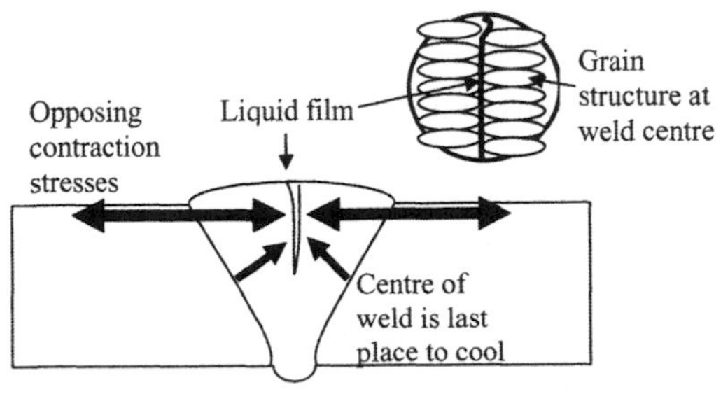

Figure 7.1 Solidification cracking

apart along the weakened centre of the weld bead (see Fig. 7.1).

Solidification cracking can be avoided by the following methods:

Adding manganese (to form manganese sulphide). The manganese combines with the liquid sulphur to form manganese sulphide (MnS), which has a melting point similar to steel. The manganese sulphide structure is spherical and forms between the solidifying weld metal grains without forming a liquid film. This maintains the cohesion between the grains and prevents a crack occurring.

Keep sulphur levels low:

- Specify low sulphur content material with a content <0.03% S.
- Keep surfaces clean of contaminants such as oils, paints and grease.
- Use welding processes that have a low dilution level of parent plate to weld metal (to reduce the levels of sulphur being introduced into the weld metal from the parent plate).
- Ensure temperature indicating crayons used close to welds do not contain sulphur.

Keep stress levels low:

- Reduce restraints to decrease restraint stress and residual stress levels.
- Keep heat inputs low.

Maintain low carbon levels. An increase to the carbon content can rapidly increase the manganese to sulphur ratio required to minimise cracking. Therefore use low carbon fillers and low carbon materials.

Solidification cracking occurs in austenitic stainless steels because the austenitic grain structure is intolerant to contaminants such as sulphur or phosphorus between the grains in the weld metal. A high level of these contaminants can cause the weld to crack.

This cracking can be avoided by making the weld metal about 5% ferrite because the ferrite structure is more accommodating to contaminants. Weld metal can be partially ferrite by the addition of a filler metal chosen using a Schaeffler diagram. It is worth noting that although austenitic stainless steel is non-magnetic the weld metal may be found to be slightly magnetic due to the slight ferrite content.

In welds with a large depth/width ratio (usually in excess of around 2:3) the centre of the weld is the last place to cool and therefore contains large columnar-shaped grains at this point. These grains impinge on each other as they form from both directions within the weld, leading to voids forming in the centre of the weld that cannot be filled with the molten filler being added from above. These voids weaken the weld metal, which is then subjected to severe contraction stresses caused by the shrinkage of the large weld metal volume, producing a centreline crack. Processes with deep penetration or large deposition rates such as SAW are particularly susceptible to this type of failure.

Reheat cracking

Reheat cracking occurs primarily in the HAZ of thick-section high strength low alloy (HSLA) steels, 300 series stainless steels and nickel-based alloys. During PWHT or elevated service temperatures intergranular cracking can occur due to stress relaxation in coarse grained regions under high restraint or residual stresses. The failure normally initiates at a stress concentration such as a notch or change in cross-section. Reheat cracking can be avoided by:

- adequate preheating to reduce the stress levels in the HAZ;
- using joint designs that require less restraint during welding in thick sections;
- removing stress concentrations caused by sharp changes in cross-section, such as sharp undercut, mechanical damage and poorly blended weld toes.

Lamellar tearing

Lamellar tearing occurs mainly in thick-section T-joints and closed corner joints in carbon and carbon manganese steels with a high sulphur content and/or high levels of restraint. It does not occur in cast or forged steels; only in rolled plate. It has the appearance of a 'steplike' crack (Fig. 7.2) and occurs in wrought (rolled) plates due to a combination of:

- contraction stresses from the cooling weld acting through the parent plate thickness plus

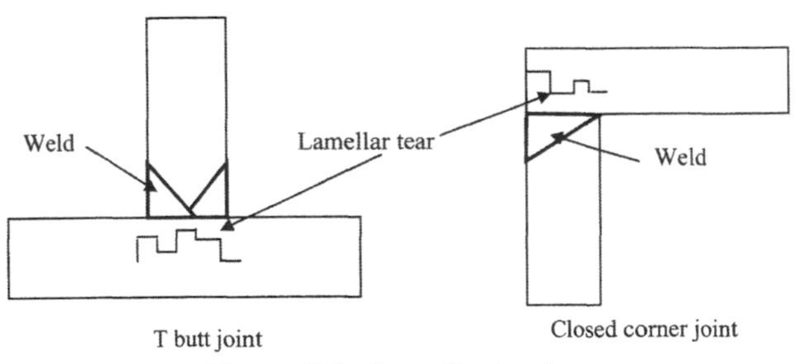

Figure 7.2 Lamellar tearing

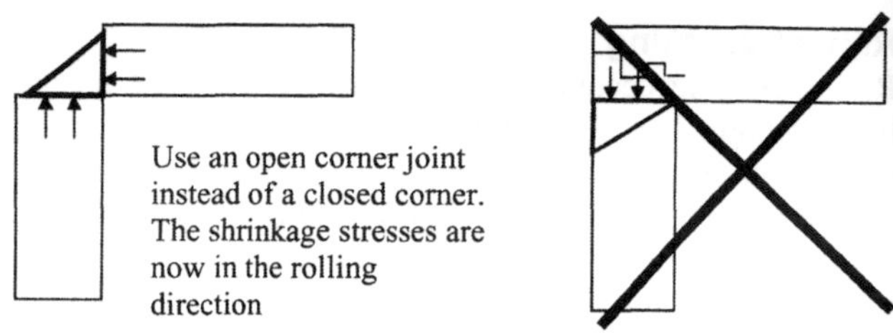

Figure 7.3 Change of joint design

- poor through-thickness ductility due to impurities in the steel (such as sulphides, sulphur, microinclusions and small laminations).

The main ways to avoid lamellar tearing occurring are to:

- *Change the joint design* (Fig. 7.3) to reduce the stress level across the rolling direction (through the thickness). Cracking does not occur when contractional stresses are in the rolling direction. You can relate this to a plank of wood where the grain is running down the length. If you apply a load in the grain direction then the wood is very strong. Apply a load across the grain, however, and the wood fractures more easily.
- *Use a buttering layer* around the joint. A buttering layer (Fig. 7.4) can be applied using a more ductile filler

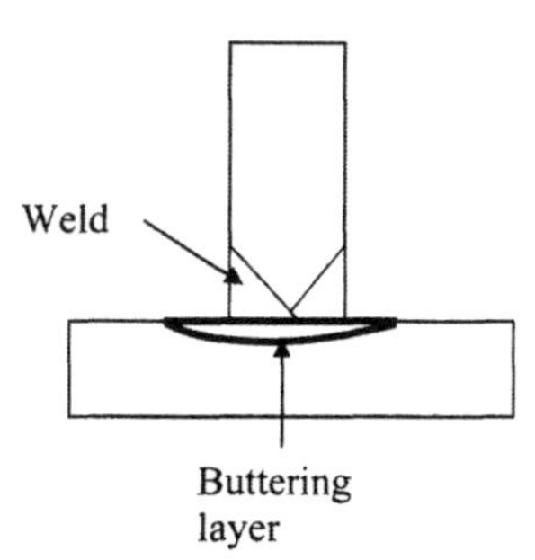

Figure 7.4 Weld buttering layer

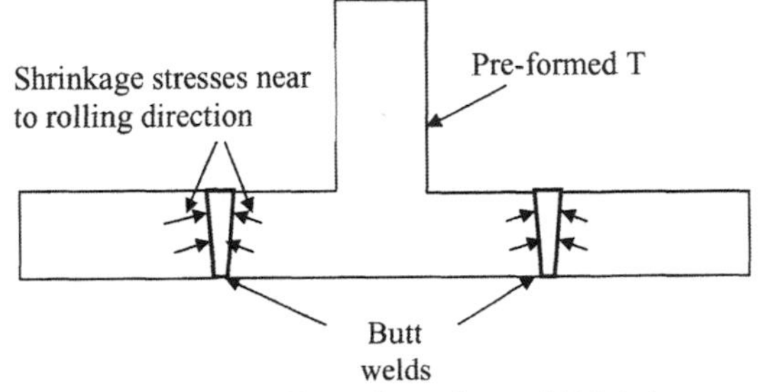

Figure 7.5 Use of pre-formed T joint

material. This allows plastic strain to be absorbed within the buttering layer and so reduce the contractional stress through the parent material thickness.

- *Reduce restraints* to reduce the restraint stress. This can be done by offsetting the parent material instead of using jigs or fixtures that reduce distortion but increase the restraint stress.
- *Use pre-formed T joints* (Fig. 7.5) which are then joined with butt welds. The butt welds can be designed to reduce the through-thickness stress by keeping most of the shrinkage stresses in the rolling direction.
- *Pre-heat* on the base material. Pre-heat will retard the cooling rate, reducing both the shrinkage stress and hardness of the base material. The reduction in hardness will decrease the risk of fracture in the material.
- Use material that has been *STRA tested.* STRA stands for short transverse reduction in area test. This is a tensile test for material that checks for ductility in the through-thickness (short transverse) direction. The greater the ductility, the less chance there is of lamellar tearing occurring. A measure of the ductility is the %*E* (% elongation) or %*A* (% reduction in area) properties of the base material.

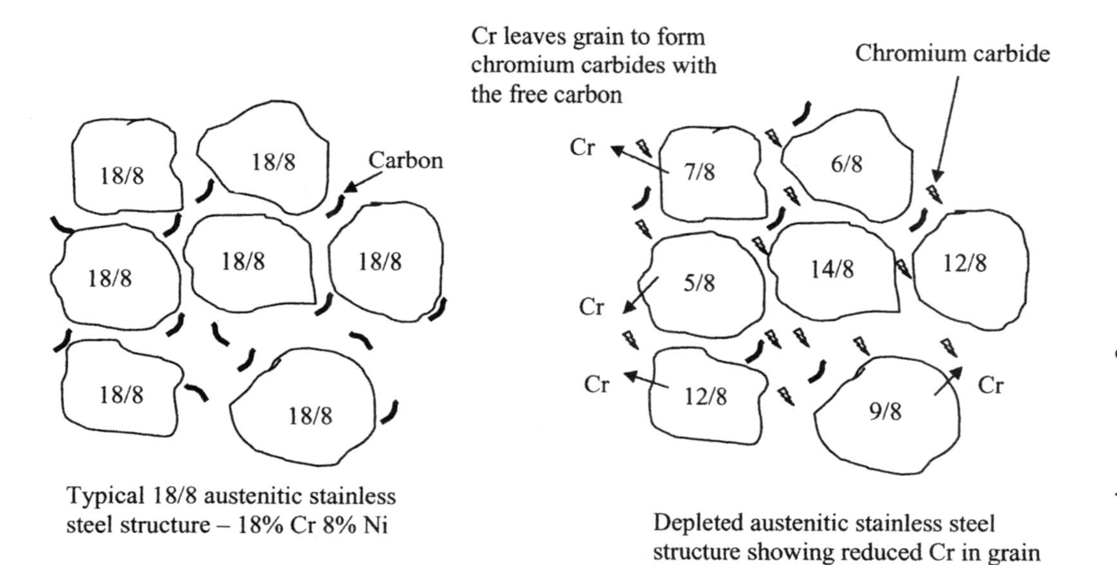

Figure 7.6 Grain depletion

Weld decay

Weld decay is a form of intergranular corrosion that occurs in the HAZ of unstabilised stainless steels. Within the temperature range of around 600—850 °C chromium comes out of solution (Fig. 7.6) to join with free carbon and form chromium carbides. The chromium was in the grain to help prevent corrosion so corrosion can now occur where it has been depleted. Once this chromium depletion occurs the depleted area is said to be *sensitised* (meaning it is susceptible to corrosion) and will corrode in the presence of an electrolyte. The critical region is usually in the HAZ parallel to the weld toes (Fig. 7.7) and once the area is sensitised, corrosion can lead to rapid failure. Weld decay can be avoided by:

- *Using low carbon grade stainless steels.* These steels have less carbon available to form chromium carbides. Reducing the carbon content also reduces the tensile strength, however, so 304L is used instead of 304 and 316L instead of 316. The L indicates low carbon grade stainless steel which may contain about 0.3% carbon rather than 0.8% carbon.
- *Using stabilised stainless steels* instead of unstabilised grades. To prevent the loss of strength associated with using low carbon grades of stainless steels, stabilised

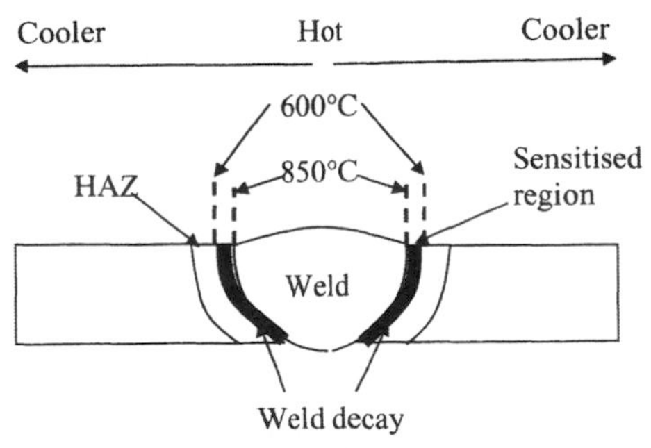

Figure 7.7 Weld decay

stainless steel grades such as 321 or 347 are used. These have stabilisers such as titanium (321) or niobium (347) added. The titanium and niobium are stronger carbide formers than chromium and form titanium carbide and niobium carbide thereby leaving the chromium in the grain.

- *Quench cooling*. Austenitic stainless steel is not hardened by quenching and is not generally susceptible to hydrogen cracking so cooling rapidly to reduce the time in the critical heat range can be done without detrimental effects.
- *Keeping heat inputs and interpass temperatures low*. These actions reduce the time that the material is held in the critical temperature range.
- *Solution heat treatment after welding*. This involves heating to around 1100 °C and quenching, which will dissolve the chromium carbides and restore the chromium to the grain. Do not carry out this procedure without consulting a metallurgist for advice because other problems can be induced in the material if you get the temperatures/timing wrong. Also, be aware that a major disadvantage of this method is the high level of distortion it causes.

Porosity

Porosity (Fig. 7.8) is the entrapment of gases (H_2, O_2, N_2, etc.) within the solidifying weld metal. Causes of porosity include:

- *Loss of gas shield.* Nitrogen and oxygen contamination result from poor gas shielding. As little as 1% air entrainment in the shielding gas will cause porosity. Draughts, leaks in the gas hose or incorrect gas flow rates are frequent causes of porosity.
- *Damp electrodes or fluxes.* Hydrogen can originate from moisture in insufficiently dried electrodes and fluxes.
- *Arc length too large.*
- *Damaged electrode flux.*
- *Moisture or contamination* on the parent material or

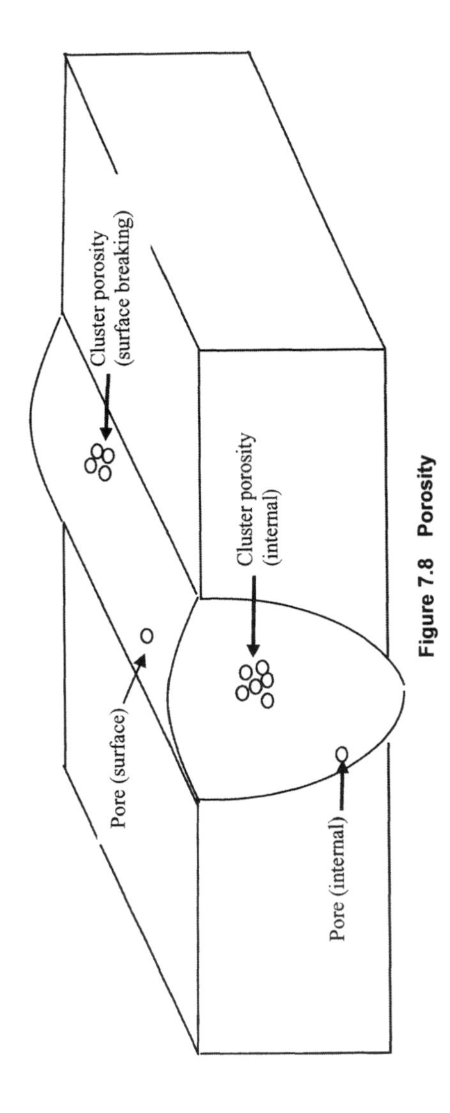

Figure 7.8 Porosity

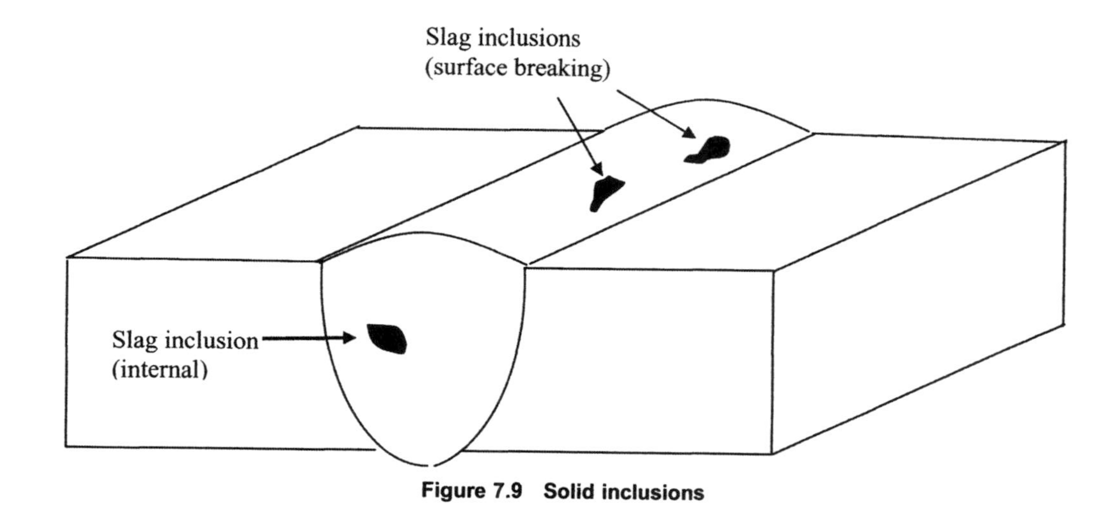

Figure 7.9 Solid inclusions

consumables. Moisture, grease, paint or oil on the material surface or filler wire are common sources of hydrogen leading to porosity.

Solid inclusions

Solid inclusions (Fig. 7.9) can be metallic (i.e. tungsten, copper, etc.) or non-metallic (i.e. slag) and are formed within the weld metal. Causes of solid inclusions include:

- inadequate cleaning of slag originating from the welding flux;
- inadequate removal of silica inclusions in ferritic steels during MAG or TIG welding;
- touching the tungsten to the weld pool during TIG welding;
- the melting of the copper contact tube into the weld pool during MIG/MAG welding.

Lack of fusion

Lack of fusion (Fig. 7.10) is weld metal not correctly fused to the parent material or the previous weld bead. Causes of lack of fusion include:

- incorrect joint preparation (narrow root gap, large root face);
- incorrect welding parameters (current too low);
- poor welder technique (incorrect electrode tilt or slope angles);
- magnetic arc blow;

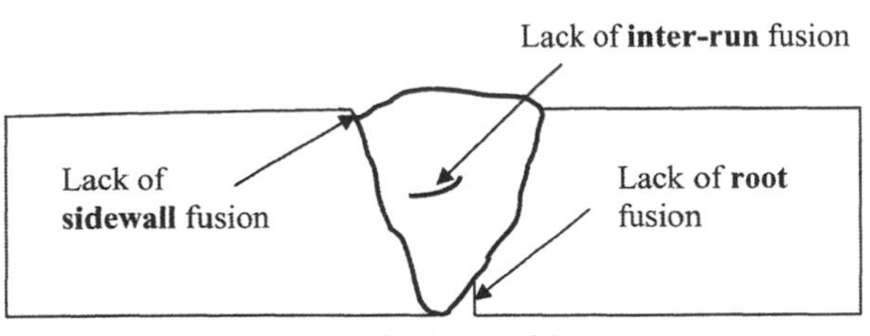

Figure 7.10 Lack of fusion

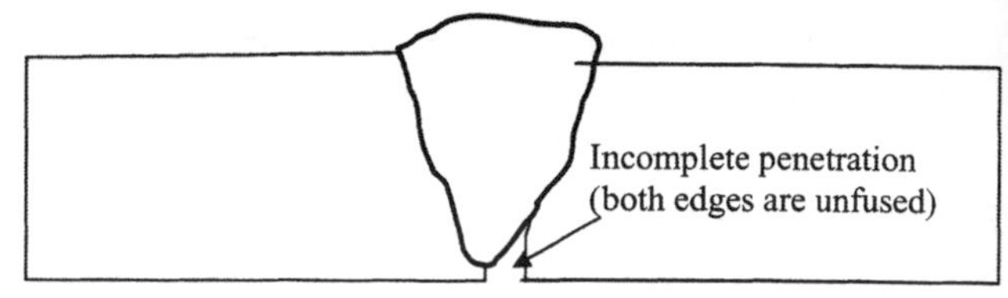

Figure 7.11 Incomplete root penetration

- poor surface cleaning.

Incomplete root penetration

Incomplete root penetration (Fig. 7.11) is where both edges of the root faces are not fused. Causes of incomplete penetration include:

- incorrect joint preparation (narrow root gap, large root face);
- incorrect welding parameters (current too low);
- poor welder technique (incorrect electrode tilt or slope angles);
- magnetic arc blow;
- poor surface cleaning.

Root concavity

Root concavity (Fig. 7.12) is a groove in the root of a butt weld, but with both edges correctly fused. It is sometimes referred to as 'suck back'. Causes of root concavity include:

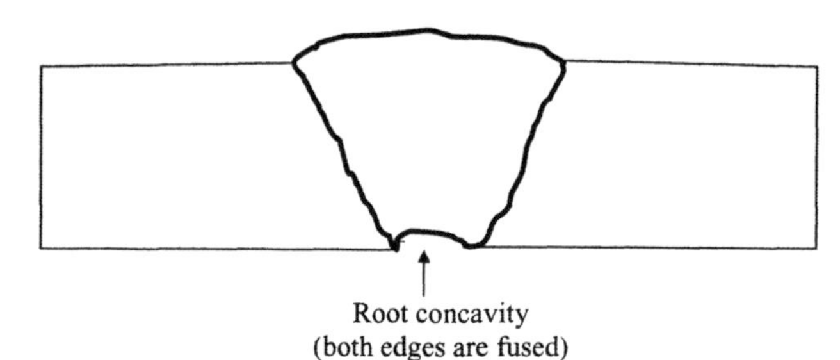

Figure 7.12 Root concavity

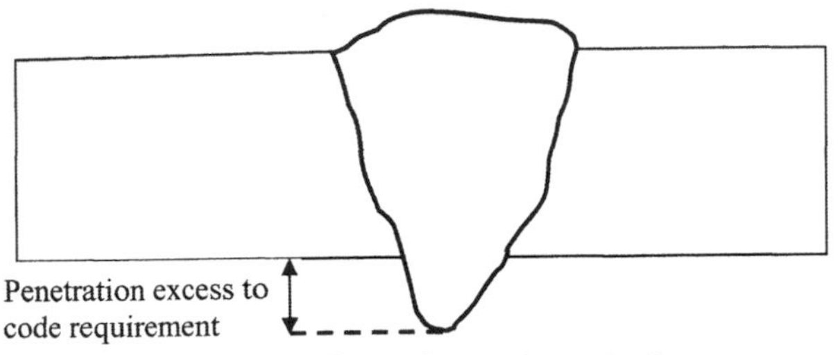

Figure 7.13 Excessive root penetration

- the root face or root gap too large;
- excessive purge pressure being applied when welding using the TIG process;
- excessive root bead grinding before the application of the second weld pass.

Excessive root penetration

A protruding penetration bead is classed as excess penetration because it is excess to requirements and does not contribute to the weld strength. If the level of root penetration is in excess of the design code acceptance criteria it is then classed as *excessive penetration* (Fig. 7.13). Do not confuse excess penetration (which may be acceptable to code requirements) with excessive penetration (which by definition is *not* acceptable to code requirements). Causes of excess/excessive penetration include:

- root faces too small;
- root gaps too large;
- excessive current leading to deeper than expected penetration;
- electrode travel speed too slow.

Overlap

Overlap (Fig. 7.14) is filler metal lying on the surface of the parent metal but not fused to it. Causes of overlap include:

- incorrect travel speed;

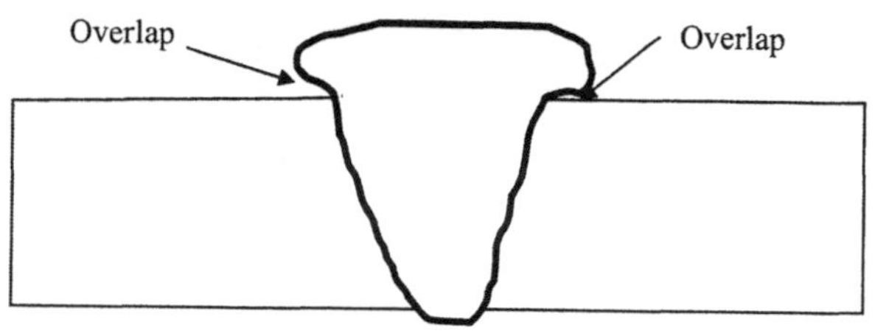

Figure 7.14 Overlap

- incorrect welding technique;
- current too low.

Underfill

Underfill (Fig. 7.15) is the term given to a joint that has not been completely filled to the parent metal surface but the edges of the joint have been fused. Causes of underfill include:

- too small an electrode being used;
- too few weld runs;
- poor welder technique.

Undercut

Undercut (Fig. 7.16) is an unfilled groove left at the toe of the weld following melting of the parent material. It can also be found in previously deposited weld metal. The problem with undercut is that it causes a reduction in material thickness and is a stress concentration from which failures such as

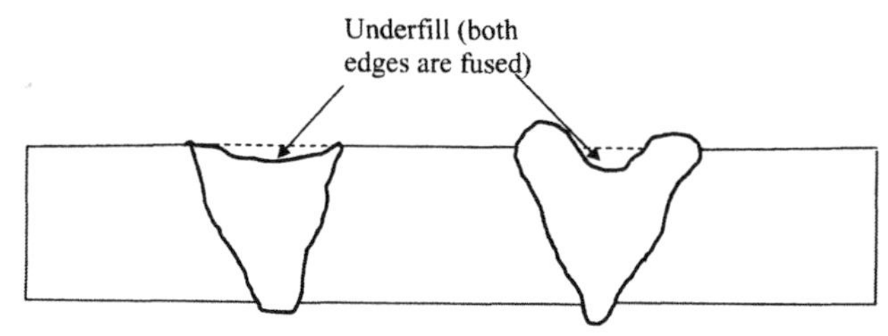

Figure 7.15 Underfill

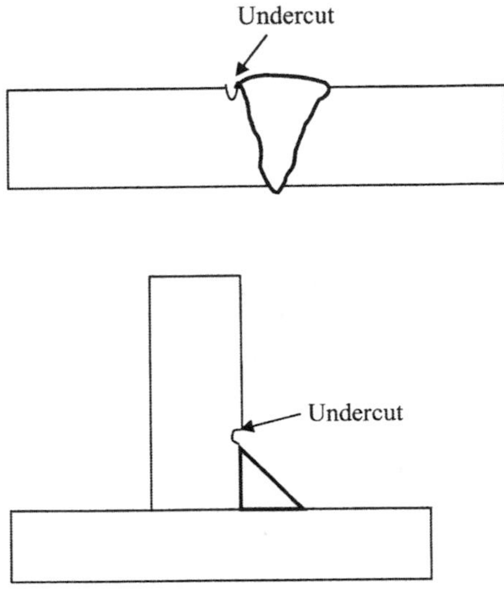

Figure 7.16 Undercut

fatigue fracture can propagate. Undercut is normally assessed by length, depth and profile (sharp or smooth) with acceptance specified by the design code. Causes of undercut include:

- excessive amps/volts;
- excessive travel speed;
- incorrect electrode angle;
- incorrect welding technique;
- electrode too large.

Crater pipe

A crater pipe results from shrinkage of the weld pool on solidification. It is usually caused by switching off a high welding current or breaking the arc resulting in the rapid solidification of a large weld pool. It can be effectively prevented when using the TIG process by progressively reducing the welding current (using a slope-down control) to reduce the weld pool size before breaking the arc and/or by adding filler to compensate for the weld pool shrinkage. Other processes may require run-off plates to be used at the end of weld runs or special techniques to be utilised to reduce the effects of breaking the arc too quickly.

Burn-through

Burn-through is a localised collapse of the weld pool during the root run indicated by excessive root penetration or an irregular cavity in the root bead. It is normally assessed on radiographs by looking at the density. Causes of burn-through include:

- excessive welding current;
- having a small or uneven root face;
- having a large or uneven root gap;
- using too slow a travel speed.

Root oxidation

Root oxidation (sometimes referred to as 'coking') is most common when welding stainless steels and leaves a blackened poorly fused root. It is caused by insufficient back purging gas.

Arc strike

An arc strike (or stray flash) is accidental arcing on to the parent material. This can lead to cracking on crack-sensitive materials due to the fast quenching of the arc strike, causing localised hardened regions. These hardened regions are susceptible to brittle fracture. They can also cause stress concentrations leading to in-service failures such as fatigue fractures. Arc strikes caused by poorly insulated cables or loose earth clamps may introduce copper or other dissimilar materials into the weldment, causing liquation cracking or other contamination problems. Arc strikes on susceptible materials require removal and PT or MT to ensure no cracking is present.

Spatter

Spatter is molten globules of consumable electrode that are ejected from the weld and quench quickly wherever they land on the weldment. They can therefore cause cracking on susceptible materials so they should be removed and then the area tested with PT or MT. Other problems caused by spatter include prevention of UT (because UT needs a smooth surface for the probes), unwanted retention of penetrant during PT and problems with paint retention. Causes of spatter include:

- excessive current;
- damp electrodes;
- surface contamination from oil, paint, moisture or grease;
- incorrect wire feed speed during MAG welding.

Magnetic arc blow

Magnetic arc blow is an uncontrolled deflection of the welding arc due to magnetism. This causes defects such as lack of root fusion or lack of sidewall fusion. Causes of magnetic arc blow include:

- deflection of the arc by the Earth's magnetic field (can occur in pipelines);
- poor position of the current return cable (the magnetic field surrounding the welding arc interacts with the current flow in the material to the current return cable and is sufficient to deflect the arc);
- residual magnetism in the material causing distortion of the magnetic field produced by the arc current.

Some methods of avoiding arc blow are:

- welding towards or away from the clamp;
- using a.c. instead of d.c.;
- demagnetising the steel before welding.

Chapter 8

Codes, Standards and Documentation

Codes and standards

Reference is often made in books (including this one) and quality documentation to 'compliance with the relevant specification, code or standard' but a common question is 'what is the difference (if any) between them?' Unfortunately there is no universally agreed definition of what specifically constitutes a code or standard but the following loose definitions are acceptable.

A **code** is generally:

- A set of rules that must be followed when providing a specific product or service.

A **standard** may:

- refer to standard procedures such as examinations and tests of materials and personnel;
- be a specification for a material or manufactured product and may either be written by companies for internal use or by national and international bodies for public use;
- be a document *referred to* by a code and contain optional or mandatory manufacturing, testing or measurement data. For example, a pressure vessel will be manufactured to a *code* using materials meeting the quality levels contained in a *standard.* In reality it is often not that simple because many standards are actually codes in all but name.

A *code of practice* is normally legally binding and contains all the rules required to design, build and test a specific product. Compliance with the ASME Boiler and Pressure Vessel (BPV) code, for example, is a legal requirement in most states of the USA and pressure vessels manufactured in accordance with Section VIII of the BPV code must comply with all aspects of the code before they can have the ASME stamp applied to them.

Conversely, compliance with the European unfired pressure vessel standard BS EN 13445 is not a legal requirement in the UK (unless it is a contractual requirement) but is the British equivalent to Section VIII of the BPV code (and is therefore realistically also a code).

The term code of practice (or code) is often used by ASME and API, or by individual companies or organisations for their own applications, but the term is not normally used by the British and European standards organisations. The term *application standard* is often used to describe a construction standard in the UK. The best way to look at this is to accept that the construction document may be called a code, an application standard, recommended practice or specification depending upon its country of origin and/or the organisation that released it and the inspector's job is to ensure the required compliance with it.

When all is said and done, a standard or code is a document of best practice that contains the lessons learned over time about the best method of manufacturing a product to meet an acceptable level of quality. Generally the higher the level of quality required then the more stringent the code/standard will be in terms of:

- design;
- the manufacturing method;
- acceptable materials;
- workmanship;
- testing requirements;
- acceptable imperfection levels.

Be aware that codes generally do not contain all the relevant data required for the design, manufacture, testing and inspection but will reference other standards and documents as required. Common construction code/application standards covering vessels, piping and pipelines are as follows:

- ASME VIII: *Rules for Construction of Pressure Vessels*
  One of the major construction codes used throughout the world is the ASME Boiler and Pressure Vessel (BPV) code. It is divided into 12 different sections, but when referring to a section within the ASME BPV code it is normal to just preface the section number with ASME. Therefore, Section V of the ASME BPV code is normally just referred to as ASME V, Section VIII as ASME VIII, Section IX as ASME IX and so on.
- BS EN 13445: *Unfired Pressure Vessels*
  This is the European harmonised standard for the construction of unfired pressure vessels. It superseded BS 5500 *Manufacture of Unfired Pressure Vessels*, which had such good technical information within it that the BSI reissued it as a Published Document (PD 5500) for reference purposes.
- ASME B31.1: *Design and construction rules for power piping*
- ASME B31.3: *Design and construction rules for process piping*
  These are the two most commonly used piping construction codes from the ASME B31 series and are used worldwide.
- BS EN 13480: *Metallic Industrial Piping*
- API 1104: *Welding of Pipelines and Related Facilities*
  This is a very common standard used throughout the world for pipeline construction.
- BS 4515-1:2009: *Specification for Welding of Steel Pipelines on Land and Offshore*
  This is the specification for carbon and carbon manganese steel pipelines.

The PED and UK PER

The European Pressure Equipment Directive (PED) is a directive that specifies the minimum essential safety requirements (ESRs) required for pressure equipment and any pressure equipment to be used within the EU must comply with these requirements by law. The PED is implemented in the UK by means of the UK Pressure Equipment Regulations (PERs).

Do not confuse the PED (a European directive) and PER (a UK regulation) with construction standards/codes. They are totally different entities. Directives and regulations are laws whereas construction codes and standards are normally taken to be 'arbiters of good practice' or guidelines. A manufacturer can choose to build a product following the requirements of a code or standard but the final product must also comply with the law (directive/regulation).

European harmonised standards

European harmonised standards are those that are considered to satisfy the relevant essential safety requirements (ESRs) specified in European product directives such as the Pressure Equipment Directive (PED). Harmonised standards contain an appendix Z, which defines which directives and ESRs the standard meets. The idea of building to a harmonised standard is that it gives a 'presumption of conformity' with any relevant European directive. Products demonstrate their compliance with relevant directives by having a CE mark affixed by the manufacturer. Not every European standard will necessarily be a harmonised standard but it is worth noting that when a European standard is released it replaces the relevant competing national standard from all the countries making up the EU.

If pressure equipment is manufactured to a non-harmonised standard it does not automatically have 'a presumption of conformity' with the PED and will therefore need to show compliance by other means. This usually entails utilising the services of a 'notified body' (NoBo) to prove conformance.

The notified body is an organisation that has been notified to the EU as having sufficient knowledge and experience to be able to ascertain compliance with the directive.

NDE standards

There are many standards dealing with all methods of NDE but it is worth being familiar with the following:

- EN 970: *Non-destructive Examination of Fusion Welds – Visual Examination*
  This European standard covers the visual examination of fusion welds in metallic materials. It is normally, but not always, performed on welds in the as-welded condition.
- EN 1206: *Non-destructive Examination of Welds – General Rules for Metallic Materials*
- ASME V: *Non-destructive Examination*
  This is actually Section V of the ASME BPV code and is also specified by other ASME and API codes for their NDE requirements.

European quality system requirements

- EN 719: *Welding Coordination – Tasks and Responsibilities*
  Welding is a special process that requires the coordination of welding operations in order to establish confidence in welding fabrication and reliable performance in service. The tasks and responsibilities of personnel need to be clearly defined. This standard identifies the quality-related responsibilities and tasks included in the coordination of welding-related activities.
- EN 729: *Quality Requirements for Welding – Fusion Welding of Metallic Materials*
  This standard contains guidelines to describe welding quality requirements suitable for application by manufacturers and is composed of the following parts:
  - Part 1: Guidelines for selection and use
  - Part 2: Comprehensive quality requirements
  - Part 3: Standard quality requirements

- Part 4: Elementary quality requirements

Welding procedure qualifications

Welding procedure qualification is carried out to prove that a welded joint meets the mechanical, metallurgical and physical properties required by a code or specification. It also enables repeatability by allowing a systematic approach to making the welded joint to be recorded and then used by any qualified welder or welding operator. The welding procedure qualification documents consist of a welding procedure specification (WPS) and a procedure qualification record (PQR). The PQR is also referred to as a welding procedure qualification record (WPQR) in the European procedure qualification standards. For the purposes of clarity this book will use the terms PQR and WPS.

Common procedure qualification standards are:

- ASME IX: *Welding and Brazing Qualifications*
  This specifically details the welding procedure qualification requirements for welding in accordance with the ASME BPV code but is also referenced by other American standard organisations including API. Although other standards will reference ASME IX they may specify additional essential variables or procedure qualification requirements.
- EN 288-2: *Specification and Approval of Welding Procedures for Metallic Materials – Part 2: Welding Procedure Specification for Arc Welding*
  EN 288-2 is the only remaining part from a series of EN 288 standards covering welding procedure qualification. The other parts have been superseded by EN ISO 15607 through to EN ISO 15614, which are standards detailing the specification and qualification of welding procedures for metallic materials with respect to:
  - general rules;
  - guidelines for a metallic materials grouping system;

- welding procedure specification for both arc and gas welding;
- qualification based on tested welding consumables;
- qualification based on previous welding experience;
- qualification by adoption of a standard welding procedure;
- qualification based on pre-production welding test;
- welding procedure test: arc and gas welding of steels and arc welding of nickel and nickel alloys.

Procedure qualification record (PQR)

The PQR is a record of the welding data and variables, which were used to weld a test coupon. It also contains the results of any NDE and/or mechanical tests carried out on the test coupon. Different qualification codes have different requirements but the PQR will contain, as a minimum, the essential variables used in the production of the test coupon. In many cases the non-essential variables will also be recorded, but this is not necessarily a mandatory requirement.

ASME IX is rather unusual in that it separates the toughness requirement of a welded joint from the other mechanical property requirements and assigns it a separate set of essential variables called *supplementary essential variables*, which are only required when impact testing for toughness is carried out and recorded on the PQR.

Figure 8.1 shows an example blank PQR form which contains the requirements recommended by ASME IX.

Note that the main procedure qualification codes (BSEN/ISO) used within Europe have different requirements to the American (ASME/API) codes and they do not fit well together. Procedures may have to be separately qualified to both codes if production welding is specified to comply with both European and American construction codes.

Welding procedure specification (WPS)

A WPS contains all the information required by the welder or welding operator to produce a production weld that meets

QW 483 PQR format

Company Name________________________

Procedure Qualification Record No. ______________ Date______

WPS No________________________

Welding Process(es)________________

Types (Manual, Automatic, Semi-Auto)________________

JOINTS (QW-402)

Groove Design of Test Coupon

(For combination qualifications, the deposited weld metal thickness that shall be recorded for each filler metal or process used)

BASE METALS (QW-403) Material spec.______ Type or Grade ______ P-No ______ to P-No______ Thickness of test coupon ______ Diameter of test coupon ______ Other ______	**POSTWELD HEAT TREATMENT (QW-407)** Temperature ______ Time ______ Other ______ **GAS (QW-408)** Percent composition Gas(es) (Mixture) Flow Rate Shielding ____ ____ ____ Trailing ____ ____ ____ Backing ____ ____ ____
FILLER METALS (QW-404) SFA Specification ______ AWS Classification ______ Filler metal F-No ______ Weld Metal Analysis A-No ______ Size of Filler Metal ______ Other ______ Weld Metal Thickness ______	**ELECTRICAL CHARACTERISTICS (QW-409)** Current ______ Polarity______ Amps ______ Volts ______ Tungsten Electrode Size ______ Other ______
POSITIONS (QW-405) Position of Groove ______ Welding Progression (Uphill, Downhill) ______ Other ______ **PREHEAT (QW-406)** Preheat Temp. ______ Interpass Temp.______ Other ______	**TECHNIQUE (QW-410)** Travel Speed ______ String or Weave Bead ______ Oscillation ______ Multiple or Single Pass (per side) ______ Single or Multiple Electrodes ______ Other ______

Figure 8.1a PQR example (ASME IX)

the required mechanical and metallurgical properties. All the required essential and non-essential variables will be recorded in the WPS forming the instructions to the welder (and also to the inspector). The WPS will contain the

QW 483 PQR (Back)

PQR No. __

TENSILE TEST (QW-150)

Specimen No	Width	Thickness	Area	Ultimate Total load lb	Ultimate Unit Stress psi	Type of Failure & Location

GUIDED- BEND TESTS (QW-160)

Type and Figure No	Result

TOUGHNESS TESTS (QW-170)

Specimen No	Notch Location	Specimen Size	Test Temp	Impact Values			Drop Weight Break (Y/N)
				Ft-lb	% Shear	Mils	

Comments ____________________

FILLET WELD TEST (QW-180)

Result – Satisfactory? : Yes_____ No ______ Penetration into Parent Metal? : Yes_____ No _____

Macro – Results ____________________

OTHER TESTS

Type of Test ____________________

Deposit Analysis ____________________

Other ____________________

Welder's Name ____________ Clock No. ____________ Stamp No._______

Tests conducted by: ______________ Laboratory Test No. ______________

We certify that the statements in this record are correct and that the tests welds were prepared, welded, and tested in accordance with the requirements of Section IX of the ASME Code.

Manufacturer ____________________

Date__________________ By ____________________

(Detail of record of tests are illustrative only and may be modified to conform to the type and number of tests required by the Code.)

Figure 8.1b PQR example (ASME IX)

qualified range of a variable based on the actual value, recorded in the PQR, which was used to weld the test coupon. The qualified range of a variable specified in the WPS is known as the 'range of approval' or 'extent of

ASME IX QW 482 WPS format

Company Name ____________ By : ____________
Welding Procedure Specification No ______ Date ______ Supporting PQR No ______
Revision No ______ Date ______

Welding Process(es) ____________ Type(s) ____________
Automatic, Manual, machine or Semi Automatic

JOINTS (QW-402)

Details

Joint Design ____________
Backing (Yes) ______ (No) ______
Backing Material (Type) ____________
Refer to both backing and retainers

☐ Metal ☐ Nonfusing Metal

☐ Nonmetallic ☐ Other

Sketches, Production Drawings, Weld Symbols or Written Description should show the general arrangement of the parts to be welded. Where applicable, the root spacing and the details of weld groove may be specified.At the option of the manufacturer, sketches may be attached to illustrate joint design, weld layers and bead sequence, eg for notch toughness procedures, for multiple process procedures etc

BASE METALS (QW-403)

P-No. ______ Group No. ______ to P-No. ______ Group No. ______
OR
Specification type and grade ____________
to
Specification type and grade ____________
OR
Chemical Analysis and Mechanical properties ____________
to
Chemical Analysis and Mechanical properties ____________

Thickness Range:
Base Metal: Groove ______ Fillet ______

Other ____________

FILLER METALS (QW-404) Each base metal-filler metal combination should be recorded individually

Spec. No (SFA) ____________
AWS No Class) ____________
F No ____________
A-No ____________
Size of filler metals ____________
Weld Metal
Thickness range
Groove ____________
Fillet ____________
Electrode-Flux (Class) ____________
Flux Trade Name ____________
Consumable Insert ____________
Other ____________

Figure 8.2a WPS example (ASME IX)

approval'. Figure 8.2 shows an example blank WPS form that contains the requirements recommended by ASME IX.

Welder qualifications

A welder is qualified by welding a test coupon in accordance with an approved welding procedure. The main purpose of

ASME IX QW 482 (Back)

WPS No ____ Rev _____

POSITIONS (QW-405)
Position(s) of Groove ______________
Welding Progression: Up ___ Down ___
Position(s) of Fillet ______________

POSTWELD HEAT TREATMENT (QW-407)
Temperature Range ______________
Time Range ______________

PREHEAT (QW-406)
Preheat Temp. Min ______________
Interpass Temp. Max ______________
Preheat Maintenance ______________
(Continuous or special heating, where applicable, should be recorded)

GAS (QW-408)

	Gas(es)	Percent composition (Mixture)	Flow Rate
Shielding			
Trailing			
Backing			

ELECTRICAL CHARACTERISTICS (QW-409)
Current AC or DC ______________ Polarity __________
Amps (Range) ______________ Volts (Range) _______
(Amps and volts range should be recorded for each electrode size,position, and thickness, etc. this information may be listed in a tabular form similar to that shown below).

Tungsten Electrode Size and Type ______________
(Pure Tungsten, 2% Thoriated, etc)
Mode of Metal Transfer for GMAW ______________
(Spray arc, short circuiting arc, etc)
Electrode Wire feed speed range ______________

TECHNIQUE (QW-410)
String or Weave Bead ______________
Orifice or Gas Cup Size ______________
Initial and Interpass Cleaning (Brushing, Grinding, etc) ______________
Method of Back Gouging ______________
Oscillation ______________
Contact Tube to Work Distance ______________
Multiple or Single Pass (per side) ______________
Multiple or Single Electrodes ______________
Travel Speed (Range) ______________
Peening ______________
Other ______________

Weld Layer(s)	Process	Filler Metal		Current		Volt range	Travel Speed Range	Other (remarks, comments, hot wire addition. technique, torch angle etc)
		Class	Diameter	Type Polarity	Amp Range			

Figure 8.2b WPS example (ASME IX)

this qualification test is to prove that the welder has the knowledge, skill and dexterity necessary to produce a weld that meets the requirements of the relevant standard or specification.

Welder qualification testing is generally the responsibility of the manufacturer or contractor but some standards require the test to be witnessed by an independent third party examiner. One of the duties of a welding inspector is to oversee a welder qualification test being carried out and ensure that it is being done in compliance with the requirements of the relevant code or specification.

A common question asked is whether a welder who welds a test coupon to qualify a new welding procedure is automatically qualified to weld that procedure (even though the coupon has in effect been welded using an unqualified procedure). If not, it means that the welder would have to weld to the procedure once it was qualified to gain the qualification. The answer is very simple; it depends on what the standard says. European standards will normally permit this but others may not.

It is important to be aware of the differences between standards in how welders are qualified. The two main standards are ASME IX and BSEN 287. Many fabricators require their welders to be qualified to both if they are required to comply with the requirements of ASME and the requirements of European directives such as the Pressure Equipment Directive (PED). The ideal situation would be to qualify welders to both standards on the same form using the same test coupon, but unfortunately this is not possible because the qualified thickness ranges, pipe diameters and the non-destructive testing and destructive testing requirements are different. It is therefore better to qualify the welder separately to each standard and apply the relevant qualification in production.

Common welder qualification standards are:

- ASME IX: *Welding and Brazing Qualifications*
  Article III contains welder qualification requirements for welding done in accordance with ASME or API requirements. Requirements may also be referenced within other standards or specifications.
- EN 287-1:2004: *Approval Testing of Welders – Fusion*

Welding of Steels
This is the welder qualification standard used within Europe for fusion welding steels in accordance with European standards.

- EN ISO 9606: *Approval Testing of Welders – Fusion Welding*
  This is the welder qualification standard used within Europe for fusion welding of materials other than steels in accordance with European standards. EN ISO 9606 is divided into Parts 2 to 5 covering aluminium, copper, nickel, titanium, zirconium and their respective alloys. There is no Part 1 because agreement could not be reached for EN 287-1 to be replaced by it.

Figure 8.3 shows the suggested welder qualification form for ASME IX while Fig. 8.4 shows the welder qualification form for EN 287-1. An interesting point to note is that the ASME IX qualification lasts indefinitely as long as the welder is involved in the use of the specified process and the manufacturer signs his qualification history to this effect every six months. This is different to the EN287-1 qualification, which also requires signing every six months but normally only lasts for two years (although it can in some circumstances be extended for a further two years).

Essential variables

Essential variables are those that will affect the mechanical or metallurgical properties of a weldment. If they are changed outside their qualified range a new procedure has to be qualified. The essential variables will be determined by the code that is being used so you do not need to guess what they are. It is also worth bearing in mind that what is an essential variable in one code may very well not be an essential variable in another code so it is important to read the code requirements closely.

The essential variables are recorded in the PQR and it is from these values that the range (or extent) of approval will

QW-484A WELDER PERFORMANCE QUALIFICATION
(See QW-301, Section IX, ASME Boiler and Pressure Vessel Code)

Welder's Name ____________ Identification No ____________

Test description

Identification of WPS followed ____________ □ test coupon □ production weld

Specification of base metals ____________ Thickness ____________

Testing Conditions and Qualification Limits

Welding variables	Actual Values	Range Qualified
Welding process(es)		
Type (ie. Manual, semi-auto) used		
Backing (metal, weld metal, double welded etc)		
□ Plate □ Pipe (enter diameter if pipe or tube)		
Base metal P- or S-Number to P- or S-Number		
Filler metal or electrode specification(s) (SFA) (info only)		
Filler metal or electrode classification(s) (info only)		
Filler metal F-Number(s)		
Consumable insert (GTAW or PAW)		
Filler type (solid/metal, flux cored/ powder)		
Deposit thickness for each process		
Process 1 ________ 3 layers minimum? □ Yes □ No		
Process 2 ________ 3 layers minimum? □ Yes □ No		
Position qualified (2G, 6G, 3F etc)		
Vertical progression (uphill or downhill)		
Type of fuel gas (OFW)		
Inert gas backing (GTAW, PAW, GMAW)		
Transfer mode(spray/globular or pulse to short circuit-GMAW)		
GTAW current type/polarity (AC, DCEP, DCEN)		

RESULTS

Visual examination of completed weld ____________

□ Bend test □ Transverse root and face (462.3a) □ Longitudinal root and face (462.3b) □ Side 462.2

□ Pipe bend specimen – corrosion resistant overlay □ Plate bend specimen – corrosion resistant overlay

□ Macro test for fusion (QW-462.5b) □ Macro test for fusion (QW-462.5e)

Type	Result	Type	Result	Type	Result

Alternative radiographic examination (QW-191) ____________

Fillet weld – fracture test (QW-180) ________ Length and percent of defects ________

Macro examination (QW-184) ______ Fillet size (in) ____x____ Concavity/convexity (in) ______

Other tests ____________

Film or specimens evaluated by ____________ Company ____________

Mechanical tests conducted by ____________ Laboratory test no. ____________

Welding supervised by ____________

We certify that the statements in this record are correct andthat the test coupons were prepared, welded and tested in accordance with the requirements of Section IX of the ASME Boiler and Pressure Vessel Code

Organisation ____________

Date ____________ By ____________

Figure 8.3 WPQ form (ASME IX)

be determined and recorded on the WPS. Again, the ranges will be specified by the code.

Non-essential variables

Non-essential variables are those that do not affect the mechanical or metallurgical properties of a weldment. If changed outside their qualified range a new procedure will

Annex A
(informative)

Welder's qualification test certificate

Designation(s): ....
....
WPS – Reference: Examiner or examining body – Reference No ....

Welder's Name:
Identification:
Method of identification:
Date and place of birth:
Employer:
Code/Testing Standard:

Photograph
(if required)

Job knowledge: Acceptable/Not tested (delete as necessary)

	Test piece	Range of qualification
Welding process(es)		
Product type (plate or pipe)		
Type of weld		
Material group(s)		
Welding consumable (Designation)		
Shielding gas		....
Auxiliaries (e.g. backing gas)		....
Material thickness (mm)		
Outside pipe diameter (mm)		
Welding position		
Weld details		

Type of qualification tests	Performed and accepted	Not tested
Visual testing		
Radiographic testing		
Fracture test		
Bend test		
Notch tensile test		
Macroscopic examination		

Name of examiner or examining body
Place, date and signature of examiner or examining body
Date of welding
Validity of qualification until:

Confirmation of the validity by employer/welding co-ordinator for the following 6 month (refer to 9.2)

Date	Signature	Position or title

Prolongation for qualification by examiner or examining body for the following 2 years (refer to 9.3)

Date	Signature	Position or title

Figure 8.4 WPQ form (EN 287-1)

not be required but the WPS has to be amended to reflect the new range before production welding takes place. Non-essential variables have to be addressed on the WPS but are not necessarily required on the PQR (again, the code will specify what is required).

PROCEDURE QUALIFICATIONS
QW-253
WELDING VARIABLES PROCEDURE SPECIFICATIONS (WPS)
Shielded Metal Arc Welding (SMAW)

Paragraph		Brief of variables	Essential	Supplementary Essential	Nonessential
QW-402 Joints	.1	Δ Groove design			X
	.4	- Backing			X
	.10	Δ Root spacing			X
	.11	± Retainers			X
QW-403 Base Metals	.5	Δ Group Number		X	
	.6	T limits impact		X	
	.7	T/t limits > 8 in (203mm)	X		
	.8	Δ T Qualified	X		
	.9	T pass >½ in (13mm)	X		
	.11	Δ P-No qualified	X		
	.13	Δ P-No 5/9/10	X		
QW-404 Filler Metals	.4	Δ F-Number	X		
	.5	Δ A-Number	X		
	.6	Δ Diameter			X
	.7	Δ Diameter > ¼ in (6mm)		X	
	.12	Δ AWS classification		X	
	.30	Δ t	X		
	.33	Δ AWS classification			X
QW-405 Positions	.1	+ Position			X
	.2	Δ Position		X	
	.3	Δ ↑↓ Vertical welding			X
QW-406 Preheat	.1	Decrease >100°F (56°C)	X		
	.2	Δ Preheat maint.			X
	.3	Increase >100°F (56°C) IP		X	
QW-407 PWHT	.1	Δ PWHT	X		
	.2	Δ PWHT (T & T range)		X	
	.4	T limits	X		
QW-409 Electrical Characteristics	.1	> Heat input		X	
	.4	Δ Current or polarity		X	X
	.8	Δ I & E range			X
QW-410 Technique	.1	Δ String/weave			X
	.5	Δ Method cleaning			X
	.6	Δ Method back gouge			X
	.9	Δ Multiple to single pass/side		X	X
	.25	Δ Manual or automatic			X
	.26	± Peening			X

Legend:
+ Addition   > Increase/greater than   ↑ Uphill   ← Forehand   Δ Change

- Deletion   < Decrease/less than   ↓ Downhill   → Backhand

Figure 8.5 Brief of variables for SMAW

Supplementary essential variables

These are only found in ASME IX and become essential variables when toughness testing is required as part of the procedure approval. ASME IX contains tables called the 'Brief of Variables' for most of the common fusion welding processes and these tables specify whether a variable is essential, non-essential or supplementary. The 'Brief of Variables' table does not fully explain what the actual variable is but gives direction to the relevant code paragraph that does give a full explanation. Figure 8.5 shows the table for the SMAW welding process.

Chapter 9

Health and Safety

Health and Safety at Work Act

The Health and Safety at Work Act (HASAWA) is a piece of legislation that states that your employer has a duty under the law to ensure, so far as is reasonably practicable, your health, safety and welfare at work. In particular, your employer must:

- Assess the risks to your health and safety.
- Make arrangements for implementing the health and safety measures identified as being necessary by the assessment.
- If there are five or more employees, record the significant findings of the risk assessment and the arrangements for health and safety measures.
- Make sure that the workplace satisfies health, safety and welfare requirements, e.g. for ventilation, temperature, lighting, and sanitary, washing and rest facilities.
- Make sure that work equipment is suitable for its intended use, so far as health and safety is concerned, and that it is properly maintained and used.
- Prevent or adequately control exposure to substances that may damage your health.
- Take precautions against danger from flammable or explosive hazards, electrical equipment, noise and radiation.
- Avoid hazardous manual handling operations, and, where they cannot be avoided, reduce the risk of injury.
- Provide health surveillance as appropriate.
- Provide free any protective clothing or equipment, where risks are not adequately controlled by other means.

As an employee you have legal duties too. They include:

- Taking reasonable care for your own health and safety and that of others who may be affected by what you do or do not do.
- Co-operating with your employer on health and safety.
- Correctly using work items provided by your employer, including personal protective equipment, in accordance with training or instructions.
- Not interfering with or misusing anything provided for your health, safety or welfare.

A welding inspector is not a Health and Safety Officer but must be aware of the dangers associated with welding and cutting operations and comply with their legal duty under site procedures and/or national legislative requirements (i.e. HASAWA) to ensure the safety of themselves and other workers. Some dangers will be common to any workplace environment and some will be specific to the welding/cutting environment. The following sections contain some typical dangers associated with an environment where welding and cutting operations take place.

Electrical safety

When welding, electrical safety practices must be followed including identifying all electrical equipment and having it tested for electrical safety on a planned basis. Equipment must have the correct insulation, be in good condition and be suitably earthed. A low voltage 110 V supply should be used where appropriate for all power tools and the correct fittings and connections must be fitted to the cables and equipment.

A duty cycle is an important consideration and refers to the amount of current that can be safely carried by a conductor in a period of time. The time base is normally ten minutes and a 60% duty cycle means that the conductor can safely carry this current for six minutes in ten and then must rest and cool for four minutes. At a 100% duty cycle the equipment can carry the current continuously. Generally 60% and 100% duty cycles are given for welding equipment.

An example would be a welding machine that can operate at 350 amps at a 60% duty cycle but only 300 amps at a 100% duty cycle.

Do not confuse a duty cycle with a welding process's operating factor. They are often confused because both are given as a percentage. A *process operating factor* is multiplied by its deposition rate in economic calculations to calculate welding costs including process down time. The down time is the time when an arc is not struck and no welding is taking place.

Some typical process operating factors are:

- MMA: 30%;
- MIG/MAG: 60% (hence the possible confusion with a duty cycle);
- MIG/MAG mechanised/automated: 90%.

Remember that the operating factor percentage refers to the percentage of time that an arc is struck.

Welding/cutting process safety

The safety considerations required for cutting and welding operations are there to ensure:

- Suitable ventilation and extraction is in place and operating correctly.
- There are no combustible materials such as oily rags in the area.
- Confined spaces are gas and fume free with gas free certificates in place.
- PPE is worn at all times.
- Permits to work are in place and complied with.
- Oil and grease is kept away from oxygen gauges and fittings (it can ignite).
- Fittings with a high copper content are not used with acetylene gas systems because explosive copper acetylyde can be formed.

Personal protective equipment (PPE)

PPE is defined in the Personal Protective Equipment at Work Regulations (1992) as 'all equipment (including clothing affording protection against the weather) which is intended to be worn or held by a person at work and which protects him against one or more risks to his health or safety'. Hearing protection and respiratory protective equipment are not covered by these regulations because other regulations apply to them, but they do need to be compatible with any other PPE provided. The main requirement of the regulations is that PPE is to be supplied and used at work wherever there are risks to health and safety that cannot be adequately controlled in other ways and that it is:

- Properly assessed before use to ensure it is suitable.
- Maintained and stored properly.
- Provided with instructions on how to use it safely.
- Used correctly by employees.

Arc welding can cause injury in many ways so the correct PPE protection is essential and includes:

- Fire retardant overalls to protect skin from burns caused by the powerful visible and ultraviolet light emitted. They also protect from spatter (globules of burning metal) which is thrown out during welding.
- Gloves to protect hands from burns from hot metal.
- A welding mask with dark filter where the mask itself can afford direct protection from fumes and spatter. The correct grade of filter will protect the eyes from the visible and ultraviolet (UV) light emitted from the welding arc. This UV light can lead to a severe eye irritation called 'arc eye' (which feels like sand has been thrown into your eyes). Any welder who has experienced arc eye will tell you how painful it is and how little they would wish to repeat the experience.
- Boots with steel toecaps to protect from falling objects.

- Goggles/face mask for eye protection when grinding or chipping.
- Gaiters and spats can give protection from burning metal or spatter falling into boots and burning feet. Otherwise, ensure overalls are outside boots.
- A leather apron may be required depending on the process used.
- A welder's hat is especially useful when performing overhead welding.

Gases and fumes

Exposure to the dangerous fumes and gases produced by welding can lead to many health problems, ranging from minor respiratory problems to major respiratory problems (such as no longer breathing at all). Metallic fumes (fumes containing metal particles) and/or gases can come from electrodes, plating or base metals used during the welding cycle. Metallic fumes produced from welding cadmium and zinc are extremely toxic and can result in death, so the importance of removing coatings or plating before welding cannot be overemphasised. The use of the correct extraction or breathing system is essential to maintain a safe welding environment.

Dangerous gases that may be produced during the welding process include ozone, nitrous oxides and phosgene (a form of mustard gas caused by the breakdown of some degreasing agents in arc light). These gases are all extremely poisonous and overexposure can result in death. Other gases used in welding cause problems by displacing air or reducing the oxygen content and so cause death by asphyxiation or suffocation. An example is the use of inert argon gas in a confined space; the argon is heavier than air and will sink to the bottom of the space and reduce the oxygen content as it slowly displaces the air and fills the space from the bottom up. Helium, on the other hand, will fill a space from the top down.

Gases such as oxygen or acetylene can build up in a space

and then ignite, causing a flash fire or explosion. Oxygen is particularly dangerous as it has no smell, but will cause a flash fire to occur if an ignition source is brought into an oxygen-rich atmosphere. In particular, if clothing is contaminated with oxygen, it will catch fire easily and burn very fiercely, resulting in severe injury. Even fire-retardant clothing will burn if contaminated with oxygen. Also, oxygen can cause explosions if used with incompatible materials. In particular, oxygen reacts explosively with oil and grease.

Most gases are stored under high pressure and gas cylinders should be stored and used in accordance with recognised practices. All fittings such as pressure regulators, flashback arrestors, flowmeters, gas hoses, etc., should be inspected at regular intervals and maintained in good condition.

COSHH and workplace exposure limits (WEL)

The Control of Substances Hazardous to Health (COSHH) Regulations require employers to prevent exposure to substances hazardous to health, if it is reasonably practicable to do so. If prevention is not reasonably practicable, the employer must adequately control exposure by putting in place measures including, in order of priority, one or more of the following:

- Use processes that minimise the amount of material used or produced, or equipment that totally encloses the process.
- Control exposure at source (e.g. local exhaust ventilation) and reduce the number of employees exposed to a minimum, the level and duration of their exposure and the quantity of hazardous substances used or produced in the workplace.
- Provide personal protective equipment (PPE), e.g. face masks, respirators, protective clothing, but only as a last resort and never as a replacement for other control measures that are required.

Examples of the effects of hazardous substances include:

- Skin irritation or dermatitis as a result of skin contact.
- Asthma as a result of developing allergy to substances used at work.
- Losing consciousness as a result of being overcome by toxic fumes.
- Cancer, which may appear long after the exposure to the chemical that caused it.

The Health and Safety Commission (HSC) in the UK has established workplace exposure limits (WEL) for a number of substances hazardous to health, including fumes or gases that may be produced through welding. These limits are intended to prevent excessive exposure to the substances by keeping exposure below a set limit. A WEL is the maximum concentration of an airborne substance, averaged over a reference period, to which employees may be exposed by inhalation. The WELs are listed in document EH40/2005: *Workplace Exposure Limits*. The reference periods used are eight hours for the long term exposure limit (LTEL) and 15 minutes for the short term exposure limit (STEL). The toxicity of a substance can be gauged by the value of the exposure limit, so applying good practice will mean actual exposures are controlled below the WEL. Advice on applying the principles can be found in the COSHH *Approved Code of Practice (ACOP)*. Be aware that the values given in Guidance Note EH40 can change annually. Table 9.1 contains examples of substances that may be present in welding under certain conditions.

Table 9.1 Workplace exposure limits (WELs)

Fume or gas	Long term exposure limit (LTEL) (8 hour reference period)	Short term exposure limit (STEL) (15 minute reference period)
Aluminium (respirable dust)	4 mg/m^3	Nil
Cadmium oxide fume	0.025 mg/m^3	0.05 mg/m^3
Iron oxide fume (as Fe)	5 mg/m^3	10 mg/m^3
Ozone	Nil	0.4 mg/m^3/0.2 ppm
Phosgene	0.08 mg/m^3/ 0.02 ppm	0.25 mg/m^3/ 0.06 ppm
Argon	No WEL value Air O_2 content to be controlled	

Lifting equipment/pressure system requirements

Many items of site welding equipment will require lifting equipment to place them into position for the site welding to take place. The correct lifting equipment and practices will therefore be required for all slinging and lifting operations. Lifting equipment is normally subject to regular inspection in accordance with a country's published regulations. In the UK the regulations governing lifting equipment and lifting operations is the *Lifting Operations and Lifting Equipment Regulations (LOLER)*, which lays down mandatory inspection periods for lifting equipment and accessories ranging from strops and shackles to cranes.

Fixed gas systems in Great Britain are subject to the requirements of the *Pressure System Safety Regulations (PSSR: 2000)*, which deal with the in-service inspection of pressure systems containing a relevant fluid. Gases above 0.5 barg are included in the relevant fluid classifications. Another set of UK regulations that welding equipment will be subject to is the *Provision and Use of Working Equipment Regulations*

(PUWER), which covers tools and equipment used in the workplace.

Hand tools and grinding machines

Hand tools must be kept in a safe and serviceable condition and only used by persons trained to use them correctly. Portable cutting and grinding equipment should have the correct disc fitted for cutting or grinding suitable for the material in use. Fixed grinding machines should only have their wheels changed by a suitably qualified and approved person and must always be used in a safe and correct manner.

Other precautions before welding starts

- Ensure persons in the area are shielded from welding arcs (use screens).
- Give a warning to persons in close proximity before striking an arc.
- Do not arc weld close to degreasing baths as UV decomposes the vapour into a toxic gas.
- Ensure firefighting equipment is in good order and close to hand.
- Ensure fire sentries (if required) are briefed and in position.

Causes of accidents

Accidents do not just happen but are usually attributable to one of the following:

- Poor communication (by the spoken or written word).
- Unfamiliarity (with the equipment, workplace or working practices).
- Just doing something wrong (because you haven't checked if it's ok to do it).
- Complacency (you reckon it will be fine – but based on what?).

Do not leave things to chance. If in doubt – check it out.

Permit-to-work

Many companies operate a written permit system for hot work. The permit details the work to be carried out, how and when it is to be done, and the precautions to be taken. A written permit system is likely to result in a higher standard of care and supervision and must be adhered to.

Tanks and drums

Flammable liquids and vapours such as petrol, diesel, fuel oil, paints, solvents, glue, lacquer and cleaning agents are found in many places of work. If welding or thermal cutting is used on a tank or drum containing a flammable material the tank or drum can explode violently. People have been killed and seriously injured by such explosions. Tanks and drums that are 'empty' usually still have residues in the bottom, and in seams and crevices. Just a teaspoon of flammable liquid in a drum can be enough to cause an explosion when heated and turned into vapour.

You must never weld or thermally cut a drum or tank that has contained or may have contained flammable material unless you know it has been made safe. If it contains flammable material, it will need thorough cleaning or inerting (see the HSE guidance note CS15: *The Cleaning and Gas Freeing of Tanks Containing Flammable Residues*). It may be safer for a specialist company to carry out the work. If in doubt, ask.

Index